D1518964

Voyaging the World's
CIVIL ENGINEERING WONDERS

Panama Canal Centennial Bridge, completed in 2004. (P&O Cruises)

VOYAGING THE WORLD'S
CIVIL ENGINEERING WONDERS

John Laverick MBE, CEng, FICE, MIStructE, FCMI,
Fellow of the Institution of Civil Engineers

The History Press

This book is dedicated to the rock, the foundation, upon which a civil engineering career was built, my dearest wife Valerie. Together we continue to enjoy the passion of that life we built together.

First published 2017

The History Press
The Mill, Brimscombe Port
Stroud, Gloucestershire, GL5 2QG
www.thehistorypress.co.uk

© John Laverick, 2017

British Library Cataloguing in Publication Data.
A catalogue record for this book is available from the British Library.

ISBN 978 07509 8436 2

Typesetting and origination by The History Press
Printed in Turkey

Cover illustrations
Front, clockwise from top: The Falkirk Wheel, Fife (British
Waterways); SS *Great Britain* 'afloat' in her building dock
in Bristol (SS Great Britain Trust); Gatun Locks, Panama
Canal, 15 April 2015 (Pablo Hidalgo-® 123RF.com). *Back*:
Telford's Pontcysyllte Aqueduct, a World Heritage Site on
the Llangollen Canal in North Wales (British Waterways).

A peaceful sunrise over Suez City, 2014.

CONTENTS

ACKNOWLEDGEMENTS

My particular gratitude to Sir William McAlpine for providing the Foreword to this book.

I would like to thank the following friends for their tremendous help in sourcing and providing many of the images included in the book: Keith Bennett, Richard Brayden, Robert Coles, Mike Crompton, Paul Lenaerts, John Minns, Tim Pyatt, Jürgen Rohweder, Doug Small, Robert Yeowell.

Thanks also to these organisations for the generous provision of information which I have been able to use in this voyage through the world of civil engineering: Arup (Simon Birkbeck), British Waterways, Broads Authority, Canal & River Trust (David Viner), Carnival UK, Cotswold Canals Trust (Clive Field), Google Earth (for general permission to reproduce in books), SS Great Britain Trust (Nick Booth), Imperial War Museum, National Waterways Museums (Doreen Davis), Nord-Ostsee Kanal [Authority] (NOK), PSA Defence Works Navy, Paddle Steamer Preservation Society (Myra Allen), Panama Canal Authority (ACP), Royal Haskoning (David Goodman), Suez Canal Authority, Thames Water, Van Oord.

Special thanks to: The Institution of Civil Engineers for linking the publication of this book with the Institution's 200th Anniversary Celebrations (1818–2018) and for providing a copy of the 'Anniversary Logo'.

Rob Luckham (my brother-in-law) who diligently read through every dot and comma of my early draft manuscript.

Triforce Solutions – for 'electronically saving the day'!

Books and Papers Read during the Preparation of this Book

The Path Between the Seas, David McCullough

Proposal for the Expansion of the Panama Canal, Panama Canal Authority 2006

Weston Mill Lake, HM Naval Base, Devonport. 'Design of Frigate Maintenance Berths', Pearce & Lohn, ICE Proceedings paper 9653 1992

Barton Broad Project Clear Water 2000, Submission to Millennium Commission, Broads Authority, September 1996

'No Easy Answers', Draft Broads Plan, Broads Authority, 1993

Parting The Desert: The Creation of the Suez Canal, Zachary Karabell

All photographs in this book have been taken by the author, unless otherwise stated.

FOREWORD

This book is based upon the lectures given by the author to entertain international audiences enjoying journeys across the oceans of the world aboard ships built for luxury travel. It contains stories of how seagoing vessels are literally able to voyage through some of the greatest works of civil engineering on the planet, and other stories of how civil engineers, over time, have changed the built and natural environment, provided defence for our country and even created some of the earliest ocean liners.

I first met John Laverick in 2000, when I was one of the judges of the British Construction Industry Awards for that year, his project having reached 'shortlisted' status in the civil engineering category. Our paths again crossed in January 2014, while travelling on board the Cunard liner *Queen Elizabeth* on the outward leg of her World Voyage in that year, a voyage that included a transit of the phenomenal Panama Canal. I attended John's programme of insights lectures and was so moved by his enthusiasm for his subject, and so impressed by the work of canal restoration volunteers featured in one of those talks that on return to the UK I joined the Wilts & Berks Canal Trust. Since that time, I have been honoured by the Trust having invited me to become its president, a position I now proudly hold.

All of my life I have been fortunate to have been involved in the exciting world of civil engineering, working with people who never say that something cannot be done on grounds of difficulty or expense; civil engineers, who are trained to be positive, will always determine how a problem might be resolved, how much it will cost and how long it will take. I share in John's vision that by providing young people with an insight into the varied works of the civil engineer, be that through the design, construction or maintenance processes, then they might be inspired to follow one of the most creative and practical career paths in the world. This book provides that insight and does so in a most entertaining and easy to read manner.

The Hon. Sir William McAlpine BT

INTRODUCTION

Through the pages of this book I invite the reader to journey with me to view civil engineering projects across the world, individual projects that have been designed to enable vessels of all sizes – ranging from ocean liners to river passenger boats, from container ships to canal barges and from large warships to submarines – to travel through them, under them or over them and to be lifted by them or be docked alongside them.

My aim in writing this book is to provide an insight into the exciting world of civil engineering, to inform those who do not know just how much civil engineers contribute to the wellbeing of society, and also to inspire those who might be considering a career in this most worthwhile of human endeavours.

Each chapter is based upon an illustrated lecture that I have presented to passengers on board ships at sea. Those lectures were designed to inform but, also, to entertain, just as this written work is intended to do.

So What is Civil Engineering?

The London-based Institution of Civil Engineers, considered by many as the premier body of professional engineering in the world, describes civil engineering as being all about helping people and shaping the world. Civil engineers design roads, railways, tunnels and bridges to provide communication between communities within countries and between countries. They build and maintain infrastructure such as ports, harbours, canals and airports that keep the commerce of the world moving. They contribute to the defence of our way of life by providing essential infrastructure to our armed forces. They keep us switched on and powered up, by supplying electricity and gas to our homes. They give us clean water and, after its first use, purify it so we can use it again. They find clever ways of recycling our waste. They restore bodies of polluted water and create new habitat for endangered species. Above all, civil engineers are creative people who solve problems.

Why Civil Engineering?

From early times through to the Middle Ages and beyond, people lived in modest dwellings. Buildings of any significance were built for military purposes; castles and defensive maritime structures stood out alone in the built environment. From the early industrial age came large mill buildings, factories, canal infrastructure and (later) railways. These were not examples of military engineering but were examples of non-military or *civil* engineering – built by civil engineers.

Why did I become a Civil Engineer?

A long career in civil engineering started for me when, at a young age, I received my first 'Meccano set' as a Christmas gift from my parents. Meccano was a model construction system consisting of reusable metal strips, plates, angle brackets and girders all joined together with real nuts and bolts. Using this, I was able to build structures and mechanisms either by following clearly worded and illustrated plans and specifications or by using my own imagination to build structures from scratch. Add to my interest in this construction medium the fact that Mr Brunel became a schoolboy hero of mine – what chance did I have not to follow the profession of civil engineering?

My Route to Professional Status

My route to professional status would today be considered somewhat unconventional and currently impossible to repeat. Known as 'the hard route' (when compared with three years at university), it consisted of working full-time for five years in an engineering office, coupled with attendance at an engineering college for one day and three evenings per week. I believe that such a system produced all-round and practical engineers. At the end of this period of office (including site training) and academic work, those following a structural engineering specialism, as I was at the time, would then join with those who had spent the previous three years at university to sit the Institution of Structural Engineers professional examination. This was, and still is, a most rigorous practical examination paper lasting seven and a half hours. Because of changing entrance qualification requirements (the degree route was to become the only route to professional status) I found myself in the very last group allowed to sit the final professional examination, having followed the so-called (and wrongly called in my opinion) 'part-time education route' – the combined office and academic path that I followed was very much full-time! I was somewhat relieved to pass the examination and, also, delighted to have been awarded the Institution of Structural Engineers' Francis Memorial Prize en route – a prize that is still awarded each year to the student with the best performance in the final year of training in the field of structural

engineering. I was duly elected as a Member of the Institution of Structural Engineers in 1972.

A few years later and, now, very much involved in the wider field of civil engineering, I recognised the benefit of seeking membership of the Institution of Civil Engineers. By this time, the only route of entry was via a university degree, and I did not have a degree. I was, however, through membership of the Institution of Structural Engineers, recognised as a 'chartered engineer' and the Chartered Engineers Institute (later to become the Engineering Council) offered to engineers in my position the opportunity to sit a transitional examination known as the 'Two Subject Test'. So, two subjects were selected for me by the Institution of Civil Engineers, a course of part-time study was followed and the exam was taken and passed. I was then in a position to make my formal submission to the Institution and attended its comprehensive interview, following which, in 1977, I was elected a Member, becoming a Fellow of the Institution of Civil Engineers fifteen years later.

My Civil Engineering Career

My working career started in the commercial ports sector. Then, following a spell with a structural engineering consultant, it developed in government service in both civilian and tri-service Ministry of Defence (MOD) areas of work before I specialised in maritime civil engineering with the Royal Navy. During the years of the Cold War, the work with which I and my colleagues were engaged had a real sense of purpose, but, afterwards (and as my Defence Works Navy Department was being privatised), I judged that this was the right time to leave the organisation, where I had reached the level of Director of Operations, and move into other areas of civil engineering.

In the entrance foyer of the last MOD office in which I was based was an honours board celebrating an impressive unbroken record of 134 years' service to the Royal Navy. Listed was the name of every senior occupier of that Devonport building from 1859 – which just happens to be the year of the death of my hero, Isambard Kingdom Brunel (1806–59) – right up until the time of my own departure in 1993. That office building has since been demolished and the site

redeveloped and, no doubt, that large mahogany board ended up in a waste skip, but I reflect with some pride that my career path allowed me to play a small part in that particular naval civil engineering story.

I then relocated from Plymouth to Norfolk to restore Barton Broad (a man-made lake), a project having a neat link back to my time with the Royal Navy. It is reputed that the Broad was the body of water on which, as a boy, Admiral Lord Nelson learned to sail.

Enjoying, now, developing facilities for small vessels (which was a real contrast to my previous works for commercial and military shipping), I continued in this vein for some years, going on to project manage the restoration of the Kennet & Avon Canal across southern England. That project was the recipient of national awards including one from The Engineering Council for 'engineering in the natural environment'. This was another tranche of work of which I am particularly proud, being able to put back into working order the grand works of civil engineer John Rennie (1761–1821).

On completion of the restoration project, I moved into general canal management with British Waterways (the then government department with responsibility for some 4,000km of canals and rivers) before finally retiring from full-time employment in 2006, but not, I hasten to add, retirement from civil engineering, which I continued for a further three years in a part-time consultancy role. In parallel with this consultancy work, I also embarked, in 2006, upon voluntary unpaid roles in river management and in canal restoration and management, the latter of which continues to this day – once a civil engineer always a civil engineer!

John Laverick
MBE CEng FICE MIStructE FCMI
Fellow of the Institution of Civil Engineers
Keevil, Wiltshire, England

1

THE PANAMA CANAL

ITS HISTORY, CONSTRUCTION AND OPERATION

Map of the Panama Canal. (Cunard; Mike Crompton)

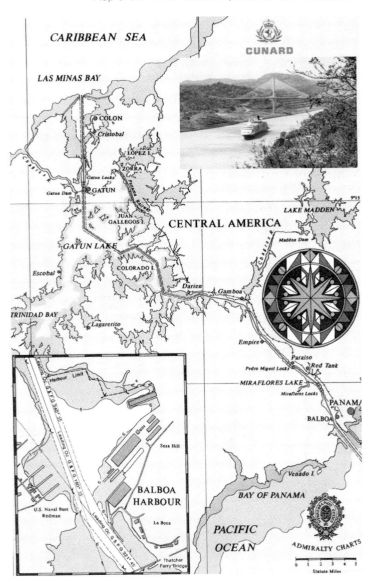

Cunard's *Queen Victoria* and a third-generation mule. (Mike Crompton)

Now, for anyone who enjoys an ocean voyage, to transit the Panama Canal is a one-of-a-kind journey that must be experienced at least once in a lifetime. To see large passenger ships and huge container ships climbing over the mountains of Panama from one ocean to the other is a truly stunning spectacle. Well, how did it all begin?

When the Spanish conquered Central America they had, over a long period, looked for a passage for their ships to navigate from the Atlantic Ocean to the Pacific Ocean. Perhaps unsurprisingly, one of the earliest proposals in modern times for a crossing of Central America came from England, the cradle of the worldwide Industrial Revolution. In the middle 1830s, Cambridge graduate Baron de Thierry, inspired by the advantages that would result from a shorter trade route between England and New Zealand, proposed a canal across Panama and secured a concession to dig it. He suggested that the whole project could be completed within three years. (Speeches given at the opening ceremony of the Panama Canal in 1914 recognised that Baron de Thierry had originally held the concession for cutting the canal, but had lost his opportunity through the dishonesty of others.)

Towards the end of his life, Civil Engineer Thomas Telford, the first president of the Institution of Civil Engineers, was considering a grand scheme for a canal crossing of Central America. The possibility of building a railway or a canal was investigated but the stupendous magnitude of such an undertaking at that time discouraged progress, and the opportunity for a British-led project faded. A railway over the same route, Telford had stated, was quite out of the question because the ground was so uneven and covered with so many leaves that the trains would not stay on the tracks. (Ironically it had been suggested that the poor performance of British trains in the 1980s had been blamed on tracks being covered with 'the wrong kind of leaves.')

France, also with expanding colonies, actually got as far as entering into a contract with the government of New Granada to build a railway across what we now know as Panama, then a district of Colombia, and a grant for this purpose was made in 1848. The many obstacles and huge sums of money required for its completion discouraged the builders to such an extent that contracts were defaulted upon within one year.

The USA, following its war with Mexico in 1846 and having come into the possession of California, not surprisingly turned its attention to the necessity of a shorter route between the eastern seaboard and its then almost inaccessible possessions in the west. The discovery of gold in California attracted a multitude of 'Forty-niners' who flocked to the new Eldorado in such numbers that steamship companies began to offer direct services between the US Atlantic coast and Pacific ports via Cape Horn. The route around Cape Horn was one of three alternative routes available at that time. A journey that today we take for granted was, in ships of the period, a long, uncomfortable and dangerous voyage of some 21,000km, which often took several months to complete. Many ships foundered, either rounding the Horn or in the severe and sudden storms of the Southern Ocean.

Another route was across the Plains of the North American continent on organised wagon trains; an arduous journey often made on horseback and again lasting many months. Countless natural hazards would be met on such a journey and there was always the constant risk of attack from the many indigenous Indian tribes living within what was then a vast wilderness.

The third route was across the narrow isthmus joining Central America to the mainland mass of South America, where the journey would usually be made on foot and through almost impenetrable disease-ridden swamps and jungle. Advertisements of the day, however, suggested that from New York one could 'take a pleasant voyage to Panama, stroll across the fifty miles of the isthmus to the Pacific shore and, after another easy sea voyage, arrive in San Francisco'. However, the trip wasn't quite like that, but, no matter, that pot of gold was just over the horizon. In reality, the journey across the isthmus was more accurately described by James Stanley Gilbert in his contemporary poem:

Beyond the Chagres River

Beyond the Chagres River
Are paths that lead to death
To the fever's deadly breezes
To malaria's poisonous breath!
Beyond the tropic foliage,
Where the alligator waits,
Are the mansions of the Devil
His original estates.

Beyond the Chagres River
Are paths fore'er unknown,
With a spider 'neath each pebble,
A scorpion 'neath each stone.
'Tis here the boa-constrictor
His fatal banquet holds,
And to his slimy bosom
His hapless guest enfolds!

Beyond the Chagres River
Lurks the cougar in his lair,
And ten hundred thousand dangers
Hide in the noxious air.
Behind the trembling leaflets,
Beneath the fallen reeds,
Are ever-present perils
Of a million different breeds!

Beyond the Chagres River
'Tis said – the story's old –
Are paths that lead to mountains
Of purest virgin gold!
But 'tis my firm conviction,
Whatever tales they tell,
That beyond the Chagres River,
All paths lead straight to hell!

So not quite the easy stroll that those early advertisements had suggested!

These, then, were the three routes to the new El Dorado. The developments that eventually took place were sparked off initially by the US Post Office, which was determined that some new way had to be found to carry the growing volume of mail between the east coast of the US and California. The Post Office believed that a Panama route was the logical bridge between the oceans. Persuaded by this argument, the US Congress immediately established two new lines of mail ships: one from New York via New Orleans to Colon located on the northern shore of the isthmus and the other from Panama City on the south shore to California. These new mail ships initially helped to cope with the vast quantities of mail and passengers, but the tremendous bottleneck for this operation was, of course, that ghastly 80km hike through the jungle. The US Post Office declared that a railway to connect the two ports was the obvious answer.

Work on a railway was started by US engineers in 1850 and was completed five years later. The 76km route from coast to coast required the construction of more than 300 bridges, the largest of which was across the Chagres River at a point where the river had been know to rise by some 14m within a few hours after heavy storms. This new bridge was actually washed away as it neared completion and had to be completely rebuilt.

In its early days, the railway was to establish many records. It is said that, at a cost of $8 million, it was, per mile, the most expensive railway ever built. With a first-class single fare of $25 in gold, it was certainly, per mile, the most expensive ride by any train in the world. During the first twelve years of its operations, the Panama Railway carried gold dust and nuggets to a value of in excess of $750 million, earning revenue of a quarter of 1 per cent on each shipment. It frequently transported over 1,500 passengers in a single day and moved the US mail and the freight of three steamship companies. In 1913, the railway carried 3 million passengers and transported 2 million tons of freight across the isthmus. At that time, it was reported to have the heaviest traffic per mile of any railway in the world. Much of the freight carried during this period was, of course, generated by the construction of the canal. Sadly, upward of 12,000 people died during the five-year construction period of the railway and, as can be imagined, disposing of so many bodies was a major problem. It is widely believed that the enterprising railway company 'pickled' many of these bodies in barrels, which they then sold to medical schools all over the world. The proceeds from this trade contributed to the building of hospital accommodation in Panama for railway employees. There is wide agreement among civil engineers that the construction of the Panama Canal would have been much more difficult, if not impossible, had a railway not been built first.

As to the beginnings of the canal, in 1870, Commander Thomas Selfridge was directed by the US Department of the Navy to manage an expedition to Central America to find a route for a canal. During the following five years, a further seven such expeditions were ordered by US President Ulysses S. Grant.

Five different routes for a canal were at one time or another under active consideration. The most northerly, which was closest to the USA and therefore able to provide the shortest sea route between the eastern seaboard and California, was across the Isthmus of Tehuantepec in southern Mexico.

Nicaragua was the place where most nineteenth-century North Americans, including President Grant, expected to see a canal built. In fact, two canals were proposed: one connecting the Atlantic Ocean to the east end of Lake Nicaragua, a large natural body of water located in the centre of the country, and the other connecting the Pacific Ocean to the west end of the lake. As early as 1811, a canal across Nicaragua had been visualised by Alexander von Humboldt, a German-born naturalist and explorer, in what was probably the first authoritative study aimed at trying to find a solution to bridging the gap between the two oceans. Humboldt proposed that such a waterway could be built much along the lines of Thomas Telford's Caledonian Canal, which had been cut right across central Scotland, then the most ambitious undertaking of its kind in the world. Like Telford's use of Loch Ness, Lake Nicaragua, Humboldt suggested, would provide a natural and limitless source of water for the canal.

Panama was judged by Humboldt to be the worst possible location for a canal, primarily because of the mountains, which he took to be three times higher than they actually are. The best that he could suggest for Panama was to build a good road for camels.

In April 1870, Commander Selfridge's expedition steamed into the Gulf of San Blas, a magnificent harbour on the Atlantic coast. As viewed from the sea, the mountains gave the appearance of dividing into a low pass and, from an earlier expedition, it was known that the distance from shore to shore at this point was less than 48km. With careful measurement, Selfridge calculated the height of the mountains to be 348m above sea level. He then reported that the San Blas route was not practicable and, if adopted, a tunnel would be required for ships to pass through. Imagine that, if you can, a modern ocean liner having to pass through a tunnel. He believed that even if locks could be built to lift ships to this 'preposterous altitude' there were no rivers available to supply the canal with water.

The Gulf of Urabá in Colombia, at the point where the isthmus joins the main land mass of South America, was also surveyed by Selfridge, but once again this was judged not to be a practicable route.

The official report filed by Selfridge at the completion of his surveys stated that any canal built should 'partake of the nature of a strait, with no locks or impediments to prolong the passage of ships, it must be a cut through at the level of the sea just like the canal at Suez, and from what was known of Central America the only feasible place for such a passage was across Panama'.

The Suez Canal had been successfully completed in 1869, and it was at an International Congress of the French Geographic Society in Paris in the summer of 1875 that Ferdinand de Lesseps, the hero of the Suez Canal, made his first declaration of interest in an inter-oceanic canal across Central America. He stated that, first, the best route had to be found and, second, a decision would have to be made as to whether or not the canal should be built at sea level or built with locks. Even before the congress had been brought to a conclusion, de Lesseps announced to the world that, in his opinion, the waterway must be a sea-level canal with no locks.

At the end of 1875, President Grant's Inter-Oceanic Canal Commission considered the five competing routes for a Central America canal. Having weighed the results of its several expeditions, the commission unanimously decided in favour of building a canal across Nicaragua. Panama received little more than a passing mention.

Within weeks of that American announcement, the French Geographic Society announced that it would now sponsor a great international congress for the purpose of evaluating the scientific considerations at stake in building a canal across Central America. The French stated that the recent work undertaken by the United States had been insufficient to enable a proper conclusion to be reached as to the most suitable route. From their subsequent evaluation, and following their success at Suez, the French were inspired to undertake what they believed to be a similar project to connect the Atlantic and Pacific oceans across Panama and not across Nicaragua, as the United States had determined. Confident that such a project could be carried out with little difficulty, in 1876 an international company was created to undertake the work. Ferdinand de Lesseps was to be the figurehead for this project and concessions were obtained from the Colombian government to dig a canal across the isthmus. de Lesseps' enthusiastic leadership, coupled with his reputation as the man who had delivered the Suez Canal, persuaded speculators and ordinary citizens from around the world to invest massively in the scheme. However, de Lesseps, despite his previous success, was not an engineer. The construction of the Suez Canal, essentially a ditch dug through a flat, sandy desert, presented few challenges. Panama, on the other hand, was to be a very different story. Whilst the mountainous spine of Central America does come to a low point at Panama, it still rises to a height of over 110m above sea level; a sea-level canal, as proposed by de Lesseps, would quite literally have required the removal of mountains. Unfortunately for this French enterprise, and for all of those who had invested in it, the project was a major financial failure. The French laboured for nearly twenty years, beginning in 1880, but disease – notably yellow fever and malaria – financial problems, fraud, the wrong engineering solution from the outset and the inadequacy of their machinery all conspired to defeat them. Their efforts cost the lives of over 20,000 men and incurred costs of 1½ trillion francs. The crash of the French company was to be recorded as the greatest business failure of the nineteenth century.

Following the crash, in 1899, the US Congress created an Isthmian Canal Commission to, once again, examine the possibilities of a Central American canal and to recommend a route. The commission first endorsed President Grant's recommended route through Nicaragua but later changed its decision in favour of a route across Panama. In January 1903 the United States offered terms to the Colombian government, which at that time controlled the Isthmus of Panama, but these were rejected as being an infringement of its national sovereignty and the compensation offered was considered inadequate. In November 1903 the Hay–Bunau-Varilla Treaty was signed between the United States and the newly established republic of Panama granting the United States exclusive rights to build and operate a canal across the isthmus. Just two weeks earlier, with the US Navy anchored in local waters, Panama had declared its independence from Colombia. The treaty made provision for financial compensation amounting to a down payment of $10 million and an annual payment of $250,000, and more importantly provided military protection to Panama guaranteeing its independence; in return, the United States secured for itself a perpetual lease on a 16km-wide strip of land for the canal. For almost 100 years, this strip of land was known as the 'Canal Zone', and was effectively administered as a part of the United States of America.

The failed French company subsequently offered all of its assets in Panama to the United States at a price of $40 million. This offer was accepted, although much of the French construction equipment and machinery was abandoned to the jungle, it being judged too small or worn out for the job in hand. The USA began work at Panama in 1904 using much bigger and more robust steam shovels and other machinery. An indication of the differing scale of the construction equipment used can be understood by comparing two preserved rail-mounted tipping trucks – the USA truck was capable of moving twenty-five times more material from the excavation site in one operation than could its earlier French counterpart. An international workforce was brought together by the United States team, with 1,000 men employed early in 1904 and rising steadily to peak at nearly 39,000 men working on the project in 1910. By far the largest contingent of this labour force was from Caribbean countries, with the largest part of that group being from Barbados. On one arrival of the steamship *Ancon* from Barbados in September 1909 there were 1,500 labourers on board:

Date	Workforce
1904 (May)	1,000
1904 (Nov)	3,500
1905	17,000
1906	23,901
1907	31,967
1908	33,170
1909	35,495
1910	38,676
1911	37,826

During the period of the construction of the canal it has been estimated that 45,000 Barbadian men and women, one quarter of the island's population, travelled to the isthmus. They sent home and brought back to Barbados much wealth, triggering a social change on the island, undermining elements of its antiquated sugar economy. In a very real sense, the era of 'Panama money' created a cultural watershed between the island's quasi-feudal past and the modern Barbados of today.

From the outset, the USA team believed that, with their huge labour force and their massive machines, they too could build a sea-level canal as envisaged by the French. But it was soon realised that this was not going to be possible. The Culebra Cut through the mountains became the principal deciding factor; a sea-level canal would require a cut so deep, and therefore so wide, that it could never be an economic proposition to construct. A revision of design to a canal with locks was made in 1906.

Key to early US success was that most of the senior civil engineers managing the project, led by John Stevens, were all railway

Cucaracha landslide. (Author's collection)

Gatun Dam hydroelectric power station.

men. It was the design and operation of a very flexible on-site railway system that enabled huge quantities of material to be efficiently moved away from the excavations. It has been calculated that if all of the material excavated for the original canal was retained on loaded railway wagons, then the train formed of those wagons would extend four times around the world.

A shattering blow to President Roosevelt, who had appointed him, occurred when John Stevens resigned as head of construction. In order that resignations could never again damage the project, the US administration decided that the military would take over the construction of the canal. Accordingly, Colonel George Washington Goethals, of the Army Corps of Engineers, was appointed to manage the project. The president himself visited the construction site in 1906 to encourage the workforce.

With good progress being made on the excavation of the main channel, attention necessarily turned to the other major structural elements of the project, the most important of which was a dam required to tame the mighty Rio Chagres and which would provide stored water to feed the great locks. The Chagres River, rising in the jungles of Panama, cut right across the proposed line of the canal before discharging all of its waters into the Atlantic Ocean, and it had been the problems associated with this river that contributed greatly to the failure of the French project. The US designs called for the dam to be built near the village of Gatun, from which it takes its name. Once dammed, the Chagres flooded huge areas of the surrounding countryside to form an artificial lake, the Gatun Lake.

With an area of around 430sq km, it became the largest man-made lake in the world. At the time of its construction, the nearly 2,500m-long Gatun Dam was the largest earth-retaining structure in the world. In a cross section it had a base 610m wide and sides sloping inwards on both sides rising more than 30m to a crest width of nearly 30m at the top.

What cannot be easily seen from a ship transiting the canal is the lower-back side of Gatun Dam. Hidden by the dam is a hydroelectric power station generating 22.5 megawatts of electricity supplying most of the operational needs of the canal. An elevated roadway traverses the dam, very useful in today's troubled times, as it provides access for patrols to monitor the security of the dam.

Gatun lock chamber with lock gates under construction. (Author's collection)

If the dam was removed, the Gatun Lake would immediately drain away and, instantly, there would be no Panama Canal.

In 1935, during a period of major improvement to the canal, another dam was built further up the Chagres River. This structure, the Madden Dam, was designed to control the flow of the river into the Gatun Lake. Once the river had flooded behind that dam, a second lake, Madden Lake, was formed. This upper lake covers an area of 65sq km and, together with the Gatun Lake, provides reservoirs of stored water for the operation of the locks and additionally provides most of the drinking water for the population of Panama. Madden Dam also has, at its base, a hydroelectric power station that produces 36 mega watts of electricity. The main dam is 274m long and rises 67m from its foundations. It was named after Martin B. Madden, an Illinois congressman who had been instrumental in obtaining the $10.6 million needed for its construction. Lake Madden has subsequently been renamed Alajuela Lake.

With the Gatun Lake established, all that remained to be done was to complete the connection of the lake to the Atlantic Ocean in the north and through the Continental Divide to the Pacific Ocean in the south – not forgetting of course the need to provide locks to allow ships to be lifted from each of the oceans up to the level of the lake. Effectively, these two separate canals, together with the lake, forms the continuous navigation known today as the Panama Canal. The formation of the Gatun Lake, capturing the headwaters of the Chagres River, altered the habit of that mighty river for all time, as it now discharges its flow into both the Atlantic and Pacific oceans and not just into the Atlantic, as it did before the canal was completed. In plan, the deep draft navigation channel through the Culebra Cut and across the Gatun Lake for most of its length actually follows the original course of the Chagres River.

The construction of the locks was the other major structural element needing to be progressed and, at their planned width of 29m, they would have been among the largest in the world. However, in 1908 the US Navy requested that the width be increased to accommodate the largest US Navy ships then in planning; consideration was also given to the size of the largest passenger ships then in construction. Chambers 34m wide with a length of 305m were finally agreed upon. The locks were built in pairs, side by side, with the central wall separating the locks having a thickness of 18m and standing in excess of 23m high. The outer walls of each pair of locks were 15m wide at the base and 3m wide at the top and the lock was filled with or emptied of water through large longitudinal culverts built into the heart of the walls. It is, of course, the dimensions of the locks that principally determine the maximum size of ship able to use the canal.

When a ship is being raised in a lock, water from the higher level transfers through the main longitudinal culverts to the lock chamber containing the ship and then, via lateral culverts, to openings built into the bottom of the lock, allowing water to fill the lock and to lift the ship. All of this transfer of water is achieved by the opening and closing of sluices, with the water moving under gravitational forces only; water is not pumped anywhere in the locking operation.

Double pairs of steel gates, 20m wide by 2m thick, are provided at each end of each lock chamber; these gates were built in situ, being fabricated from steel plate. The height of each pair of gates varies from 14–25m, depending on their location along the waterway. Miraflores lower-lock gates are the tallest, having to accommodate a 5.5m variation in Pacific Ocean tide levels. There are thirty-six separate leaves in all, weighing between 350 and 660 tons each.

With the main construction of Gatun Locks completed, in September 1913 the seagoing tug *Gatun*, an Atlantic entrance working tug used for hauling barges, had the honour of making the first trial lockage. On 7 January 1914, the floating crane *Alexander La Valley* was the first vessel to actually make a complete transit of the canal. Elsewhere in the world, on 1 August 1914, Germany declared war on Russia. Then, on 3 August, it declared war on France and, the next day, invaded neutral Belgium. Britain responded by declaring war on Germany. It was, therefore, no surprise that the official opening of the Panama Canal on 15 August 1914 lacked any meaningful international ceremony and was simply marked by the passage of the steamship *Ancon* from one end of the canal to the other.

A sectional diagram of the canal indicates the way that the waterway slices through the mountains of Panama. From the Atlantic Ocean, on the right-hand side of the diagram, the three steps of Gatun Locks lift ships up to the level of the Gatun Lake. On the left-hand side of the diagram, the single step Pedro Miguel Lock lowers ships from the Culebra Cut, which is at the same level as the lake, down to the small Miraflores Lake. Finally the two steps of Miraflores Locks lower ships down to the Pacific Ocean.

So what has the canal achieved for world shipping? The principal trade routes that benefit today are those from the US east coast to Asia and from Europe to the west coast of the US and Canada. The ocean voyage from New York to San Francisco has been reduced from 21,000km to 8,300km, a considerable saving of both time and

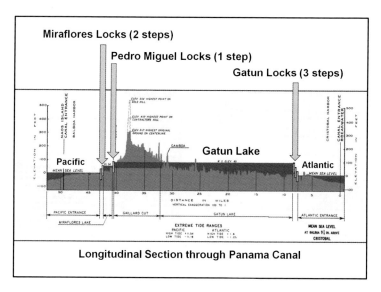

Longitudinal Section through Panama Canal

Cross section through the central divide of Panama. (Author; Panama Canal Commission)

fuel and the reason why shipping companies can justify paying the tolls charged to use the canal.

Canal tolls are decided upon by the Panama Canal Authority (ACP – Autoridad del Canal de Panamá) and are based on vessel type, size and cargo carried. For container ships, a toll of $74 (2014) was charged for each Twenty-Foot Equivalent Unit (TEU) that the ship is rated to carry. (TEU is the international standard 'Twenty Foot Equivalent Unit', the standard container that we are used to seeing singly on the back of lorries on our roads or in multiple numbers on passing trains and in their thousands on passing ships). An additional toll of $8 (2014) was charged for each loaded container. In January 2015, the Canal Authority announced that it was considering a new toll structure, which would include a carrier loyalty programme.

A Panamax ship (a ship specifically designed so that it would just fit into the locks) was the largest size of ship able to transit the canal, until the opening of the expansion project locks in 2016. Each of these ships could carry up to 4,400 containers. If you do the maths, the toll for such a fully loaded vessel for a one-way pas-

Gatun Locks three-step staircase.

The 'narrow' French canal.

sage of the canal is a little over a third of a million US dollars. For passenger ships of over 30,000 gross tons, tolls are based on the maximum passenger capacity and the ship's net tonnage. The Canal Authority have adopted their own 'Panama Canal Universal Measurement System' to determine the net tonnage used in these calculations. The ratio of net tonnage divided by the maximum passenger capacity is key to the toll calculation. If less than thirty-three, then the cruise ship will be charged on a per berth basis, which in 2014 was $134 for an occupied berth and $108 for an unoccupied berth. Where that ratio is greater than thirty-three, then tolls are charged on the basis of net tonnage only. The 2014 rate averaged $4.53 per net ton. The lowest toll ever charged for a transit was 36 cents paid by Richard Halliburton, who in 1928 swam the length of the canal, taking ten days to complete his transit. It is always very difficult to find out exactly what a particular ship pays because of special deals and 'commercial-in-confidence' issues, but it was suggested by Southampton in 2015 that the Cunard Line paid $400,000 for *Queen Elizabeth*'s nine-hour-long one-way transit of the 80km-long waterway.

So what do ship passengers get to see and experience for the proportion of their fare that they pay for a transit? Well, it's a fantastic day out and, as stated at the beginning of this chapter, to transit the Panama Canal is a one-of-a-kind journey that must be experienced at least once in a lifetime. Travelling from the Atlantic to the Pacific, a ship, when somewhere outside of the low-level rubble breakwater marking the entrance to the canal, will enter a dredged channel and the canal pilot will be welcomed on board. The Panama Canal is the only navigation in the world where a ship's captain has to hand over to the pilot almost total responsibility for the navigation of the vessel during the transit but not the overall responsibility for the safety of the vessel, which always remains with her master. After passing through the narrow entrance of the breakwater, Panama's second city Colon will be seen off the port beam, at this point the ship will still be some 10km from the first locks.

As the Gatun Locks come into view ahead, what remains of the 'French Canal' – a venture started with so much hope and enthusiasm by Ferdinand de Lesseps towards the end of the nineteenth century – will be clearly seen on the starboard side. The differing

Third-generation Mule No. 130 assisting to keep the ship parallel to the lock side.

Preserved example of a first-generation mule.

scale of the French Canal and the present waterway will be readily appreciated. On the slow approach to one side or the other of the central lead-in jetty of the Gatun Locks, a swinging signal arrow right at the end of the jetty will be seen, which used to give instruction to approaching ships as to which of the two side-by-side locks the ship must enter. These arrows are now maintained as historic artefacts, as instruction is now passed by radio.

Waiting in line on the central jetty to secure to the ship in order to assist her through the locks will be a line of tractor locomotives (powered by hydroelectricity from the dams) known as 'mules'. Once secured to these mules, the ship will move slowly forward towards the lock to be greeted on its other side by 'hobblers' – two men in a small boat who row a pulling line out to the ship from the side pier. Various higher-tech methods have been tried over the years but this remains the most efficient way to get a pulling line hoisted aboard the ship. Connected to the pulling line are wire hawsers, each of which is connected to another tractor locomotive waiting on the side pier. Once secured to the appropriate number of mules (for most passenger ships a minimum of six will be provided: two to the port bow; two to the starboard bow; and one on each side at

the stern), the approach will then be made to the first pair of lock gates, which will remain closed until the water level in the first lock chamber has been lowered to the same level as that of the Atlantic Ocean. Once these levels have been equalised, the gates will be opened, allowing the ship to proceed slowly into the first chamber and up to the next pair of gates that separate chamber one from chamber two. Once stopped in this first chamber, and held fast by the mules, the gates will firmly close behind the ship and sluices will be opened to allow water from the second lock chamber above and ahead to flow under gravity through the underground culverts to flood up through the floor of the first chamber, and so physically lift the ship until the water level in the first and second lock chambers becomes equal. Each lock requires approximately 200 million litres of freshwater to lift a ship.

The wire hawsers connecting the mule to the ship are often seen to be at approximately right angles to the ship. At this angle the mule would not be able to pull the ship forward, but that is not its principal function. By calling for a reeling in of the cable by the winch of one mule by a few inches and a corresponding reeling out of the cable by the mule on the other side of the lock, the canal pilot,

Lock Operations Control Building at Gatun Locks.

Bridge of the Americas. (P&O Cruises)

who will be on the ship's bridge, will precisely align the ship for its passage through the locks. Forward motion of the ship is achieved by rotation of the ship's propellers in the normal way. Mules, when required to do so, can provide an effective means of arresting this forward motion by moving fast towards the rear of the ship. The mules in use today are the third generation and were manufactured by the Mitsubishi Corporation at a cost of some $2 million each. These 290hp units weigh in at 50 tons and were first introduced in 1997 to gradually replace all of the second-generation mules.

Each of the second-generation mules had completely replaced all of the original mules by 1964. The first-generation mules weighed 43 tons and were fitted with a single cable drum on the top of the vehicle between driving cabs located at each end. Built by General Electric, with many coming from their factory in New York, they served the canal well for fifty years.

Once the water level in adjacent lock chambers has reached equilibrium, the gates separating the lock chambers will be opened for the ship to move forward into the next lock chamber. Two sets of gates, located close to one another, are provided at each end of the lock for added safety in case one of them might be impacted by a

moving ship. At the top of the Gatun flight, the ship will enter the third lock chamber to be lifted a final 9m up to the level of Gatun Lake. Once inside the third lock chamber, and with the ship held fast by the mules, the lock gates by which she will have entered will be closed behind the ship, forming a watertight seal. Sluices will then be opened, allowing water from Gatun Lake to start to fill the lock chamber. Looking over the side rail at this time, water will be seen bubbling up around the ship as millions of litres gush in, causing the ship to rise within the lock. Throughout this process, the pilot on the bridge will be giving instructions by radio for the mule drivers to take up slack on their cables. The lock takes about eight minutes to fill and to lift the ship to the level of Gatun Lake. Once the water in the lock chamber is at the same level as that in the lake, the sluices are closed and the upper gates opened, allowing the ship to leave the lock chamber and proceed to the point where the mules will be detached. The ship will now have been lifted a total of 26m from the mean sea level of the Atlantic Ocean up to the level of Gatun Lake.

A distinctive red-roofed control building is located at the highest point on the central wall of this and the other two lock flights. These buildings contain the main control room for each flight from which

P&O's *Artemis* in Pedro Miguel Locks, 2009. (P&O Cruises)

the operation of all sluices, lock gates and other ancillary equipment is controlled.

The navigation of Gatun Lake will start as the ship leaves the top of the three-step Gatun Locks. On the starboard side, just beyond the locks, Gatun Dam comes into view – the structure that alone maintains the water level within the lake providing all of the water for the canal operation. Entering into the wide expanse of the lake, marked navigation channels lead ships to set various courses among hundreds of islands as they navigate towards the Gaillard or Culebra Cut and beyond to the Pacific Locks. These islands are the tops of hills flooded by the rising waters of the artificial lake following completion of Gatun Dam back in 1912. Navigation channels across the lake are wide enough to allow Panamax ships to pass one another.

Eventually the wide expanse of the lake is left behind and, just before the Culebra Cut is entered off the port side of the ship, Gamboa Port will be seen – the main maintenance base for the canal. Much equipment is based at the port, including the huge floating crane *Titan* and the maintenance dredgers that can often be seen out working on the canal maintaining minimum navigation depths within the buoyed channel. These vessels are suction dredgers, essentially very large vacuum cleaners that suck a mixture of sediment and water from the bottom of the canal and pump the mixture along a floating pipeline to receptor lagoons located ashore. Just beyond the dockside, the main Panama Railway track is located and local ferries also cross the canal at this point. Beyond Gamboa, also on the port side, is the point where the wide upper Chagres River enters the canal beneath a combined road and railway bridge. This is only a single-track railway, but it is a very busy line, conveying both freight and passenger trains travelling in both directions between Panama City and Colon.

On completion of the canal, the fluvial flow of the Chagres River was altered for all time by effectively being split at this location, with half of its volume still discharging towards the Atlantic Ocean while the other half was then able to discharge via the Culebra Cut into the Pacific Ocean.

Much evidence of the work undertaken for the expansion project widening of the canal will be seen throughout the length of the Culebra Cut (see images in Chapter 2). Towards the end of this narrowest part of the canal, the waterway is crossed by the Centennial Bridge, completed in 2004. The 320m-long cable-stayed main span clears the water by 80m. The bridge has a total length of 1,052m and is supported by two towers, each 184m high. The road deck carries the six-lane Pan America Highway and the whole structure has been designed to withstand earthquakes, which are frequently recorded in the canal area.

After descending through the single-step Pedro Miguel Locks, the bottom gates will open, allowing the ship to enter the small Pedro Miguel Lake for her short transit towards the two-step Miraflores Locks. On approaching the upper chamber, the gates will open to allow entry into the first of two chambers that will lower the ship down to Pacific Ocean level. Unlike at the Atlantic end of the canal, where the three steps necessary to raise and lower ships were built as a single three-step staircase structure, here at the Pacific end of the waterway the three steps were separated out into two structures in order to avoid a geological fault crossing Miraflores Lake, which is located somewhere between the two structures. Watching a ship's progress through the Miraflores Locks, there will likely be parties of tourists receiving commentary from local guides on the balconies of a modern purpose-built lockside visitor centre.

On exiting the two-step structure, the ship, together with 200 million litres of water carried down from Miraflores Lake, will be discharged into the Pacific Ocean. The ship will have been lowered in this final pair of locks by somewhere between 13m and 20m, the actual drop depending on the Pacific tide level at the time of the transit. Before opening any of the lock gates, each of which provides a pedestrian walkway across the chamber for canal workers, handrails located on top of the gates are automatically folded down so that the gates fit neatly into recesses in the lock wall.

After departing from Miraflores Locks, the ship will then pass right through the centre of the busy Port of Balboa and its cargo quays and ship-repair facilities, with glimpses of the skyscrapers of Panama City away to port as she continues along the sea-level channel towards the Pacific Ocean to pass beneath the penultimate major structure of the waterway, the Bridge of the Americas, which was originally known as Thatcher Ferry Bridge when it was completed in 1962 at a cost of $20 million. The bridge was, until the opening of the Centennial Bridge in 2004, the only fixed bridge connecting the land masses of North and South America. Passing under the bridge, which has a total length of just over 1km, it is easy to see that with a clearance of 60m under the main span at high Pacific tide level, this structure imposes its own restriction on the size of ships that might use the canal.

Beyond the Bridge of the Americas on the port side, the mainland is linked to Naos Island by a long, continuous causeway, the last major canal structure to be seen on an Atlantic to Pacific transit. This causeway was built as part of the original canal infrastructure to help to control siltation of the deep-water channel leading to the ocean. When clear of Naos Island, the canal pilot will hand back full authority to the ship's captain and disembark onto the pilot boat. At this location, many ships will be seen at anchor, awaiting their turn to transit the canal back towards the Atlantic Ocean.

What is also worthy of note is the tremendous variety of wildlife that might be seen during a transit. Apart from a reported 546 species of exotic birds within the canal watershed, there are said to be seventy species of amphibians and 112 species of reptile in the area – on every transit I have made, I have always been lucky enough to spot from the ship's rail at least one or two large crocodiles either swimming alongside the ship or soaking up the heat of the sun at the edge of the waterway.

To complete this part of the canal story, on 7 September 1977 in Washington DC, President Jimmy Carter and the commander of Panama's National Guard, General Omar Torrijos, signed the Torrijos–Carter Treaties. These treaties abrogated the Hay–Bunau-Varilla Treaty of 1903 and guaranteed that Panama would gain control of the canal at the end of the century. Twenty-two years later in December 1999, and true to the words of that treaty, a ceremony at which Panama gained full control of the waterway from the US took place on one of the mules. Heads of state from several Latin American countries and King Juan Carlos of Spain were present for the ceremony but the notable absentee was President Bill Clinton; the highest-ranking US government representative present was Secretary of Transportation Rodney Slater. At the very last moment, Washington also informed the Panamanian government that neither Vice President Gore nor US Secretary of State Madeleine Albright would be attending. The reasons for Washington's virtual boycott of the handover ceremony are several, involving both US domestic political considerations and its strategic policy in Latin America. The canal had been a sensitive point in Republican politics ever since transfer to Panamanian control had been confirmed by Democratic President Carter. 'We built it. We paid for it. It's ours and we're going to keep it,' was the rallying cry of Ronald Reagan, who challenged Gerald Ford in the 1976 Republican primaries. Reagan later made Jimmy Carter's 'surrender' of the canal a campaign theme in his successful run for the presidency in 1980. Hopefully the politics surrounding the canal have now entered a quieter phase; ACP, the Panamanian authority that now manages the canal, has certainly demonstrated its ability to do so effectively.

The Panama Canal story does not end here. A series of major contracts were drawn up and awarded by ACP over the following decade and a half to expand the size of the canal so that it would be able to accommodate much larger ships.

2

THE PANAMA CANAL EXPANSION PROJECT

A Panamax ship just fits into a lock with 600mm to spare on each side.

Ships the size of *Queen Elizabeth 2* had no problem navigating the as-built canal, nor did *Queen Elizabeth*, *Queen Victoria* or *Arcadia*. Until recently, most modern passenger ships had been specifically designed so that they were just able to transit the canal. Ships of such a size are termed 'panamax', meaning that they are vessels of the maximum size *permitted* on the waterway, effectively the maximum size that the canal infrastructure can safely accommodate.

The ability of a passenger ship to transit the canal gives maximum flexibility to those whose job it is to plan the ever increasing variety of voyages on offer each year. Each of the canal's as-built lock chambers measures 33.5m wide and is 305m long. The maximum dimensions of a ship permitted to transit the canal are those with a beam or width of 32m and with a length of 294.3m. It is often a very tight fit – I have travelled through the canal on ships that have not only collected scrapes from their passage through the locks but have also taken away some souvenir concrete from the lock side! However, many passenger ships on the high seas today, including

Expansion project lock under construction, April 2012. (ACP)

Queen Mary 2 and *Azura*, were, due to their size, unable to transit the Panama Canal before the completion of the expansion project. Perhaps of even more significance for the operators of the waterway, container ships, the lifeblood of the canal's business, were increasingly being built to dimensions too great to use it. In fact, less than two decades after the opening of the canal, the world's shipbuilding industry produced the first vessel incapable of transiting the canal because of her size. On 29 October 1932 the 79,280-ton French luxury liner *Normandie* was launched at St-Nazaire. She was 20m too long and 3.6m too broad to be permitted to use the waterway. Two years later on 26 September 1934, *Queen Mary* was launched from John Brown's shipyard on Clydebank, Scotland. She too was greater in size than the locks could accommodate and was actually 10cm wider than *Normandie*, a seemingly insignificant difference, but very important at a time when size really did matter with the dimensions of transatlantic liners, as passengers wanted to travel on the largest and fastest ships available. Year on year there has been an increase in the number of panamax-size vessels using the canal. The percentage of total vessels transiting the canal with a beam in excess of 30m was 23 per cent in 1990, rising steadily to 45 per cent of all vessels by 2005.

Just like any system dependent on mechanical devices, the Panama Canal has a finite capacity determined by the operational times and cycles of its infrastructure. With most canals the principal limiting operational items of infrastructure are the locks. The Panama Canal is no different, and it is the Pedro Miguel Locks that are the system's main bottleneck, as these are the only ones providing a 'single lift' of a transiting vessel – the Miraflores Locks provide a 'double lift' and the Gatun Locks a 'triple lift'. When you consider that the time taken for a ship to slow down to enter any of the locks, and the time taken to attach and then detach the mule locomotives and then get up to normal cruising speed, is the same, whether the ship undergoes a single lift or a multiple lift, then you will see

that it is less efficient for all of this time to have been spent just to achieve a single lift of the ship. Whilst available water depth in the lock chambers varies, it is the south sill of the Pedro Miguel Locks that is the shallowest and, therefore, another factor contributing to these locks being the bottleneck structure for the waterway.

The ACP, the operators of the canal, forecasted back in 2006 that the waterway would reach its maximum sustainable capacity somewhere between the years 2009 and 2012 and that, once this capacity had been reached, it would no longer be possible for the waterway to continue to handle demand for growth. The authority knew that such a fall-off in service provision would rapidly lead to a decline in the competitiveness of the route, with the inevitable consequence that trade would likely move elsewhere.

Since the 1930s, various studies have been undertaken, and all had agreed that the most effective and efficient solution to enhance the capacity of the canal would be to construct a third lane of locks of larger dimension than the original two lanes. In 1939, driven by the US Navy's proposed Montana-class battleship, which would have had a length of 281m and a beam of 37m, work was actually started to construct a third lane of locks, both to increase the ability of the canal to transit larger ships and to make the operation less vulnerable to total closure that might follow from an enemy air or naval attack. The proposed dimensions of these locks were 365m long by 43m wide with a navigable depth of 14m. However, the development of the Montana-class battleship was suspended in May 1942 in favour of more urgent requirements for new aircraft carriers and anti-submarine vessels. By July 1943, when it was clear that battleships were no longer the dominant element of sea power, the Montana-class project was cancelled. But even before the Second World War, aircraft carriers were already of a size to be just able to pass through the canal, due to their overhanging island structures bristling with antennae being able to squeeze past the balconies of the lock-side control buildings. All work on the US third lane of locks project was abandoned in 1942 when the US entered the Second World War.

In the 1980s, a tripartite commission composed of the governments of Panama, Japan and the US came to the same conclusions as the earlier study. More recent studies undertaken by ACP led, in 2006, to the publication of its master plan for the future of the canal, which also confirmed that the addition of a third lane of larger locks was the most suitable, profitable and environmentally responsible way to increase the capacity of the canal. ACP's analysis of the types of vessels using the canal during the ten-year period of their survey, which ended in 2005, showed a massive increase of container traffic and, as already mentioned, container traffic is the lifeblood of the canal's business. Their analysis also showed an increase in the number of car-carrying ships and a fall-off in the transit of dry bulk cargoes, while the level of transit of liquid bulk cargo ships and passenger ships had remained fairly static throughout the survey period.

Adding to ACP's concerns was the outcome of a worldwide survey in 2006 of all container ships that were either on order or being built. This survey revealed that whilst just under a quarter of all new capacity was to be provided by small ships and just under a third by panamax-sized ships, a massive 50 per cent of all new container ship capacity was to be provided by ships of 'post-panamax' size. Post-panamax was the term then used to describe vessels of a size too large to transit the original canal.

In 2004, a market share analysis was made of the Asia to United States eastern seaboard container route corridor, then the busiest in the world. This analysis demonstrated that the Panama Canal had a 38 per cent share of traffic whilst the competing United States internal rail and road networks, linking Pacific ports to east coast markets, had a 61 per cent share of the traffic. The Suez Canal, a longer route but one already able to transit post-panamax vessels, had at the time a 1 per cent market share. ACP concluded that although there were advantages to shipping companies using the shorter sea route offered by the Panama Canal, the route was becoming less efficient, as companies were unable to upgrade to larger post-panamax ships. They were concerned that any capacity shortfalls on the Panama Canal would be taken up by the Suez Canal, as the USA cross-country routes were already operating at maximum capacity. The 1 per cent Suez Canal share of traffic grew rapidly over the years following the market share analysis.

Some canal traffic is even carried by the Panama Railway, which runs parallel to the canal for much of its route. Although the railway is quite efficient at moving containers on its double-stacked trains between post-panamax ships berthed at either end of the canal, the arrangement is not, however, the first choice for ship owners because of the necessity of the double handling of containers. A greater number of ships is also required to provide a regular interval freight liner service across the world. It would, of course, be much more efficient if those larger ships berthed at either end of the canal could transit through the waterway from one ocean to the other. The railway has even been used to lighten the load of some panamax ships that would not otherwise have been able to transit right through the canal due to their excessive draft. Some containers were taken off at one end of the canal, transported by train to the other end, where they were reloaded onto the ship. This process reduced the draft of ships, allowing them to pass over the shallow sill of Pedro Miguel Lock and to navigate some of the shallower channels across Gatun Lake.

The Maersk Line shipping company has for some time been building ships even larger than post-panamax size. The *Estelle Maersk*, when launched in 2006, was then one of the eight largest container ships ever built. Weighing in at 152,000 gross-registered tons, with a length of 397m and a beam of 56m, she is too large even for the expanded Panama Canal. Larger ships can produce other efficiencies; Maersk claimed that a coat of environmentally safe silicone paint applied to the immersed hull of this ship reduced

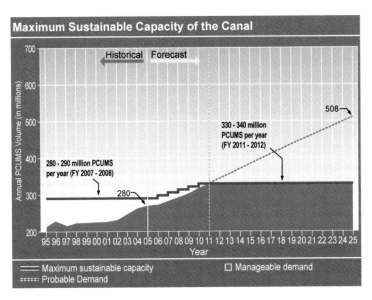

The maximum sustainable capacity of the original canal, from the 'Proposal for the Expansion of the Panama Canal', ACP April 2006. (ACP)

underwater drag, saving up to 1,200 tons of fuel per year, and that when this was combined with operating below 10 knots, whenever possible, the production of greenhouse gas emissions was around 25 per cent less than that produced by other ships of comparable size.

Other ships that will never use the Expanded Panama Canal are those of the class of ship represented by the massive $1.5 billion *Oasis Of The Seas*, a ship that has 2,700 cabins, accommodating 6,300 passengers who are served by 2,100 crew members. She measures 360m in length, has a beam of 60.5m and, with an air draft of 72m, she would not be able to navigate under the fixed bridges across the canal. She is rated five times larger than *Titanic*, which back in 1908, when the dimensions of the original Panama Locks were being considered, was one of the factors that contributed to the finally chosen dimensions adopted for those locks; the other factor was the projected size of future USA warships.

Although the market share of container movements through the Panama Canal for that key Asia–USA route corridor had steadily grown to 38 per cent by 2004, ACP was aware that the maximum sustainable capacity of the canal would soon be reached, beyond which there would be no room for any further growth. They therefore developed a strong case for the expansion of the waterway, which they stated would drive up the national economy and improve the quality of life for all Panamanians. New jobs, they said, would be created in the construction phase and, on completion, the operation of the expanded canal business would require increasing numbers of people to operate it. Just as with the building of the original canal, the new expansion contracts would enhance employment opportunities for labourers, heavy-duty equipment operators, technicians, specialists and professionals of many disciplines, including project management, surveying, design and computing, building supervision, inspection, finance, purchasing and security.

The maximum sustainable capacity of the existing canal is indicated on ACP's graph, above, by the blue line. In their 2006 forecast,

ACP determined that the demand for an expanded canal, the dotted line on the graph, would exceed available capacity at the end of 2011 at the point where the two lines cross. However, their forecast was made at a time when the world economy was still booming but the follow-on worldwide recession caused a reported downward blip of tonnage carried of between 5 and 8 per cent at the end of 2009. Modifying the ACP graph to include that downward blip perhaps suggests that demand would not actually outstrip available capacity until 2014/2015, by which time it had been hoped that the new expanded canal would have been in full operation.

Also note on the graph a stepped increase of capacity from 2006 through to 2011. This increase in capacity was achieved through a variety of provisions to squeeze that last ounce of efficiency out of the present operation. Throughout the life of the canal, improvements have been undertaken on a regular basis. In 1936, the Madden Dam was completed to increase water storage capacity for lock operations and to improve flood control of the Chagres River, which flows into the canal. Lighting projects in 1964 and

1977 increased the capacity of the canal by allowing night-time passage of vessels. The phased replacement of the first generation of mule locomotives started in 1964, improving reliability and reducing lockage times. Between 1957 and 1971 the Gaillard Cut, the narrowest part of the canal, was widened from 300ft (91½m) to 500ft (152m). This carefully planned work was achieved whilst keeping the canal open for navigation. In the 1970s, all navigational channels were deepened. From the 1980s, through to the present day, further improvement works have included another widening of the Gaillard Cut from 500ft (152m) to 630ft (192m). The second generation of mules were replaced by the third generation, these being more powerful and more efficient locomotives. The total number of mules was also increased because larger ships require up to eight mules for each locking operation. All railway tracks upon which the mules run, and which are built into the top of the lock walls, have been upgraded and replaced. The canal tug fleet has been modernised to permit the faster and safer handling of ships. Short-term mooring bays were added to some narrow channels to better manage the passage of a greater number of larger ships. All of these improvements increased the sustainable capacity of the canal but nothing short of the provision of an additional lane of larger locks would allow this capacity to be increased any further.

Lack of capacity also affected customers in another way: almost 20 per cent of users requesting booked transit slots between 2004 and 2006 could not be provided with them and had to queue on a first-come-first-served basis, with significantly lengthy waiting times. This was another clear indicator that the canal was operating close to its maximum sustainable capacity, and was obviously bad news for container ship owners trying to run regular Freightliner schedules.

As nearly every employment area within Panama is related to canal activity, it was thought essential that the canal regained its competitiveness and its capacity to capture growing worldwide tonnage demands. All of this, it was hoped, would be achieved through the Expanded Canal Project.

In summary, the three main objectives of the Panama Canal expansion project, remembering that the Canal is Panama's main economic activity, were to achieve long-term sustainability and growth for the canal's contributions to the Panamanian economy; to maintain the canal's competitiveness; and to increase the canal's capacity to capture growing worldwide tonnage demands by being able to provide appropriate levels of service. ACP forecasted that within ten years from the start of the 'Three Lane Operation' its contributions to the Panamanian National Treasury would more than double. They predicted an annual contribution of around $1,500 million in year one to in excess of $4,000 million by year ten (2025).

So, what were the components of the expansion project, a project known locally as the 'third set of locks' project? It was, of course, planned down to the smallest detail but, in essence, it was basically composed of four major elements:

The construction of two 'post-panamax' flights of locks, one at the Atlantic end, and the other at the Pacific end of the waterway (each flight comprised a three-step staircase lock together with water-saving basins).

The construction of entirely new access channels linking the new locks to the existing navigation channels.

The widening and deepening of all existing navigation channels, in the Atlantic Ocean approach, within Gatun Lake, through the Culebra Cut and within the ocean approaches at the Pacific Ocean end of the waterway.

The raising of the surface-water level of Gatun Lake, thus increasing its storage capacity.

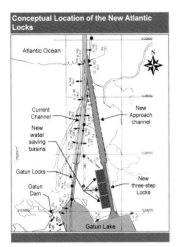

(Left) Proposed layout of the New Atlantic Locks, from the 'Proposal for the Expansion of the Panama Canal', ACP April 2006. (ACP)
(Right) Proposed layout of the New Pacific Locks, from the 'Proposal for the Expansion of the Panama Canal', ACP April 2006. (ACP)

Work in progress on the third fixed crossing, January 2016.

Many surveys had to be undertaken long before any designs were committed to paper. Just as at the start of the original construction more than 100 years before, there were many hazards with which surveyors had to deal. Perhaps not the diseases that Ferdinand de Lesseps and John Stevens had to deal with, diseases that were eventually overcome by the pioneering work of US Surgeon General William C. Gorgas, but significant hazards none the less (on every transit that I have made, I have always been lucky enough to spot crocodiles at some point along the waterway). The gathered survey information was then fed into the design process and the completed designs informed the works programmes. ACP's published programme showed most of the individual components of the project were expected to have been completed during 2014. However, due to contract delays, some of these components were still in construction up until the middle of 2016.

At the Atlantic end of the waterway, the new single three-step staircase flight of locks and its associated water-saving basins has been constructed right alongside Gatun Lake, and is accessed by a new entrance directly from the lake. The locks, now named the

Agua Clara Locks, are connected to the Atlantic dredged entrance channel by a new 3km-long approach channel. This new channel was excavated over 218m wide and deep enough to allow 'new-panamax' (a term latterly adopted by ACP) vessels to navigate in a single direction at any time, no matter what the state of the Atlantic tide level. The construction of these new locks and the access channel used a significant part of the excavations abandoned by the USA at the start of the Second World War. When I viewed the state of progress at this entrance in January 2014, the year the expanded waterway was due to be opened, this was a programme date that was clearly not going to be achieved. Based on annual visits to the canal and observing work in progress, I had been forecasting to my shipboard audiences for some time that completion of the work was likely to fall behind programme.

As part of the ACP master plan, the technical, environmental and economic viability of another fixed-link crossing at the Atlantic end of the canal, such as a tunnel or a bridge, had been evaluated but not finalised. It was, however, widely expected that construction of such a crossing would begin before completion of the expansion

Pedro Miguel Lock looking towards the Culebra Cut, January 2004.

Queen Victoria in Pedro Miguel Lock, January 2016. Note how the hills on the far side of the new 6km-long access channel have been cut back in terraces. (Mike Crompton)

project. The crossing would include water and electrical and communication provisions to facilitate development of the land area to the west side of the canal. In 2012, Vinci Construction was awarded a contract to build a third fixed crossing of the waterway. The twin-tower cable-stayed bridge is located 2km north of Gatun Locks.

At the Pacific end of the waterway, in order to connect the new lock staircase, now known as the 'Cocoli Lock Flight', to the existing navigation, two new access channels were excavated. A short southern channel connects the existing Pacific sea-level channel to the locks and a 6km-long northern channel connects the locks to the Culebra Cut. This arrangement allows 'new-panamax' ships to bypass the small Miraflores Lake. The new channels will be over 218m wide and deep enough to allow 'new-panamax' vessels to navigate in a single direction at any time, no matter what the state of the Pacific tide level. The three-step staircase flight of locks is designed to move vessels from Pacific sea level to the level of Gatun Lake at one location. Each of the three lock chambers will have alongside it

three associated water-saving basins. Just like the existing locks, the new locks are filled and emptied of water entirely by gravity, without the use of pumps. The existing two lanes of waterway through Pedro Miguel and Miraflores locks, working together with the new super locks, now forms the new three-lane regime that was originally planned to be in operation by the end of 2014, a date that was chosen to celebrate the centenary of the original opening of the canal.

Early in the construction phase, an artist's impression was created illustrating a 'new-panamax' ship entering the new 6km-long access channel, adjacent to Pedro Miguel Locks, on route to the new Pacific Locks. In January 2004 there was no sign of that channel (to be constructed to the left of the lock). The hill at the middle-top of the left-hand photograph above, which lay in the path of those excavations, would need to be removed in order to progress the works. It was in fact removed between two of my subsequent annual visits to the canal, a significant achievement for such a large volume of material.

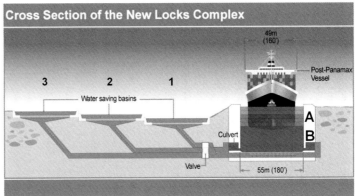

Above: Cross section through the new locks and water-saving basins, from the 'Proposal for the Expansion of the Panama Canal', ACP April 2006. (ACP)

Left: The 9m-high retaining dam between the original channel and the new channel (on the right) just below Pedro Miguel Locks.

Photographed in January 2016, that access channel now sweeps past *Queen Victoria* as she manoeuvres through Pedro Miguel Locks but, as can be seen in the photograph opposite, the channel was not quite ready for its first post-panamax ship, as final dredging was still in progress. Compare this photograph to the photograph on page 24, which shows *Artemis* in the same lock in 2009.

Pedro Miguel Locks lowers the Pacific-bound channel of the original canal by 9m; the new channel, which runs alongside for some distance, is held 9m above by a separating retaining dam.

Now let us consider the new lock structures. Each of the three lock chambers that make up the three-step staircase at both the Atlantic end and the Pacific end of the waterway is 427m long by 55m wide and 18.3m deep. The cantilever lock-side walls were built in short lengths, with gaps left between them (these gaps were later filled). This 'hit and miss' construction process helps to reduce cracking of the finished work due to the effects of concrete shrinkage. Huge longitudinal culverts, which now deliver water to the locks, were built into the bases of the walls. Each lock chamber has three associated water-saving basins running parallel to the lock and stepping up away from it. The purpose of these basins, as their name suggests, is to save water on each lock operation – precious freshwater from Gatun Lake that, without these basins, would be wasted to the ocean in a far greater quantity than is the case with the basins in operation. This idea is not new and was first used on canals in England from the very early days of the Canal Age. The basins are connected to the adjacent lock chamber by valve-controlled underground culverts. Studies undertaken by consultant Delft Hydraulic on behalf of ACP determined that the use of three water-saving basins per chamber was the optimum number and offered the highest water saving in relation to construction costs and to the impact on lockage times. The consultants also determined that this design would not adversely affect the water quality of Gatun Lake, which apart from supplying all of the water required for the operation of the locks is also the principal drinking water storage reservoir for most of Panama's population.

Four new lock gates await removal down into the new lock chambers, January 2014.

Without the basins, as a ship descends from Gatun Lake to the ocean, it would normally take the volume of water 'A' with it step-by-step to waste all of that water into the ocean. Volume 'B' reduces as each step down is taken but some will remain at the lower level in order to float the ship. Imagine volume 'A' being made up of separate slices of water. As the lock empties to lower the ship, the 'top slice' of water from the lock will discharge to fill water-saving basin No. 3. The 'second slice' of water will discharge to fill basin No. 2 and the 'third slice' will discharge to fill basin No. 1. Volume 'A' will thus have literally been put to one side to be used again. In the next lock operation, the lock will once again need to be filled, either to raise a ship from the lower level or to bring the lock back to the same level as the lake ready for another ship to descend. Without the water-saving basins, all of this water would come from the precious freshwater storage of the lake, but now 60 per cent of the water needed to refill the lock will flow under gravity from the water-saving basins and only 40 per cent will be taken from the lake. By this means, the designers of the huge new locks calculated that they would actually use 7 per cent *less* water per transit than do the original smaller locks.

Each lock chamber in the new staircase is separated by rolling lock gates, instead of mitre gates, as used on the original locks. The gates slide back into recesses at right angles to the lock, allowing ships to pass. Each recess will function as a small dry dock, which, when pumped dry with the gate in the recess, will allow maintenance of the gate without the need to remove it from the lock chamber. The need to drain the original locks to remove the mitre gates for maintenance causes significant interruption to shipping operations. It should also be remembered that mitre gates necessitate the lock chamber being built longer than that required for a ship, since when open the gates themselves take up space along the lock wall. The sixteen new Panama gates were fabricated in Trieste, Italy, and conveyed to Panama by an ocean-going heavy-lift ship

that, on arrival, moved astern to be secured to a purpose-built pier onto which the gates were rolled before being moved to storage areas until required. The gates arrived in batches of four, the last batch having arrived for storage on site in December 2014. When required for fitting, each was moved on multi-wheeled bogies along purpose-built access roads that led down into the lock chambers before being finally manoeuvred into the appropriate gate recess.

Perhaps surprisingly, given the success of mules in guiding ships through the canal locks over the past 100 years, ACP decided that tugboats alone would be used for the positioning of vessels within the new locks. They had calculated that mules, if used, would need to have been larger and heavier than those currently used and, as a consequence, a longer lockage time together with higher operational and maintenance costs would be incurred. They argued that the loadings imposed on the new lock walls would have increased the construction costs of the locks and the time needed to build them. The back of the new walls, as built, slope inwards towards the top, and the width at the top would have had to be greatly increased to support the rail track and designed with the strength to withstand the additional weights and forces imposed by the larger mule locomotives. I have always wondered whether the no mule decision was the right one, and I know that my thoughts were shared by many of the canal pilots, many of whom I have conversed with throughout the years of construction during the many transits I have undertaken.

Most of the work on the third major element of the Expansion Contract, the deepening and widening of all existing navigation channels, was undertaken by wet excavation using cutter suction dredgers. This work included the deepening of Gaillard Cut and Gatun Lake navigation channels by 1.2m. The navigation channels across Gatun Lake have also been widened to no less than 280m in the straight sections and to 366m in the turns. These dimensions allow 'new-panamax' ships to pass one another in the lake.

Artemis steams past a hill that had been terraced 100 years earlier during the canal's original construction. (P&O Cruises)

A cutter-suction dredger at work widening and deepening the Culebra Cut near to Centennial Bridge. Note that the hills alongside the channel have been terraced to allow this work to take place.

Stepping back a hillside right at the water's edge, March 2012.

A typical cutter suction dredger has a long arm, which is lowered down beneath a catamaran double hull to the required depth of the channel. At the end of the arm is a vicious rotating cutting head that fragments bottom material, which is then pumped up from the head in a pipe that is continuous through the hull of the dredger and extends out along the surface of the water, where it is intermittently supported on floats. The delivery end of this pipe would be located at prepared dumping lagoons on the shore into which the dredged material is discharged. Where the shore is too distant for a continuous pipeline, the material is pumped into bottom-opening barges and transported to designated deposit areas either at sea or in selected locations within Gatun Lake.

Whilst nearly all of this work out in the open water was undertaken by dredgers, within the narrow confines of the Gaillard Cut a whole series of operations was utilised in order to widen the navigation. In places, whole hillsides have been moved back from the edge of the channel. This has been a major operation, necessitating the removal of millions of tons of material. The widening processes used were broadly similar to the processes used during the original construction of the canal, where the hillsides were cut back away from the edge of the channel into a series of terraces.

One such hill, excavated and stepped back over 100 years ago, has almost stood the test of time but if you look closely you will see the plated ends of scores of rock anchor bolts, used to stabilise the face of the rock in order to prevent it from collapsing into the narrow navigation channel. When viewed from the other side of the canal and compared to the size of a passing ship, then the enormity of these operations can perhaps, be better appreciated. The ship is P&O Cruises former cruise liner *Artemis*.

Where hills were required to be moved and stepped back for the expansion work, this was achieved using large 360-degree earth-moving machines. The excavated material was then transported in fleets of dumper trucks along temporary haul roads cut into the local countryside to selected dumping grounds located as near to the excavation sites as possible. Mechanical shovels then managed the material into its final resting place. Compare these modern methods of excavation and spoil removal to that employed during the original construction, when giant steam excavators cut away the hillside and long trains of railway trucks were used to convey the excavated materials to the dumping grounds.

Once the hills had been stepped back from the channel by the dry excavation process, explosives were placed in thousands of drill

Steam excavators loading a train with spoil for transportation to the disposal grounds. (Author's collection)

A drilling rig in position above a shallow area at the edge of a widened channel.

holes lined with white plastic tubes on the leveled shore area. In the shallow strip of water just off of the canal bank, specialist drilling rigs were used. These rigs are held in position by four 'spud legs' – the black posts – which are independently lowered to the uneven bed of the canal to create a stable drilling platform, just like four unequal legs on a table. As soon as the platform had been stabilised, the four yellow drill towers prepared further blast holes to be filled with explosives, and then some lucky individual had the enviable task of pushing a plunger to blast the rock into small fragments. These fragments then lay just below the surface of the water and very close to the navigation channel, which needed to be buoyed off to prevent ships venturing too close until the material had been removed. Heavy grab dredgers were used to scoop up this material, which was loaded into barges tied up alongside the dredgers. These barges were constructed in two halves, hinged about the centre line and, once filled, they were towed to selected areas where the locking pins fastening the two halves of the hull were removed, allowing

the hull to split down its full length, consigning the transported rubble to the bottom of the lake or to the seabed. Alternatively if the fragmented rock was of suitable quality, then it was transported to other parts of the expansion project where it could be offloaded and used as fill material, or used in the construction of the temporary haul roads at the tipping sites.

Many disposal sites were identified in the initial surveys and later brought into use. In the northern area, seven waterborne disposal sites were identified within Gatun Lake and two more in the Atlantic approaches. Three land-based sites were identified alongside the narrow approaches to Gatun Locks. In the south, two main waterborne sites were identified in the Pacific – one inshore and one out in deep water. Eleven of the twelve land-based disposal locations were chosen because of their close proximity to the Culebra Cut, from where a huge volume of excavated material was generated.

The raising of the operational water level of Gatun Lake by 450mm was the fourth main element of the expansion project.

A grab dredger filling a bottom-opening barge in the Culebra Cut.

Spill weir at Pedro Miguel Locks.

When added to the additional volume of stored water that would already have accrued from the widening and deepening of the navigational channels, this additional overall depth greatly increased the stored water capacity of the lake; ½m of water over an area of 430sq km is an awful lot of water. The raising of the lake was principally achieved by the lifting of the sill height of Gatun Dam but in addition the sill height of the spill weirs adjacent to Pedro Miguel and Gatun Top Lock also had to be raised. The total additional storage was calculated to provide an extra 625 million litres of water per day. This was sufficient for a further 1,100 lockages per year and, in addition, the amount of water available for public consumption was increased in line with projected demands. The water level in the canal maintenance depot at Gamboa sits at lake level, the raising of which reduced the freeboard or edge depth of the quay faces that had to be built up for use by small vessels and to prevent localised flooding. This was another area where suitable excavated material was used.

The combined effect of raising the height of the lake and deepening the navigation channels means that for a given size of ship and maintaining a constant under keel clearance, then the operational margin of the lake water level is increased by a factor of three from 0.8m to 2.4m. This means that, with the new water management regime in place, then the maximum level of water in the lake, i.e. when the dam is just overtopping, can fall by up to 2.4m and still provide sufficient depth for safe passage of 'new-panamax' ships. This increase of operational margin allows for, among other things, the vagaries of rainfall patterns that are so important in these times of climate change.

ACP's original cost estimate for the project was $5.25 *billion*. When compared to what was paid to rescue bankers during the period of construction, this might be seen as a fairly moderate sum, representing, I would suggest, extremely good value for money, particularly when one considers that this new infrastructure will be providing financial returns to Panama for at least the next 100

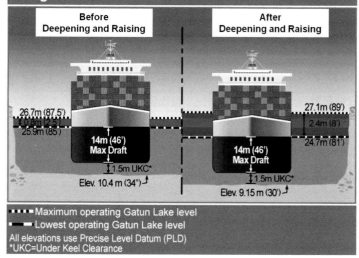

The effect of raising water levels combined with the deepening of channels, from the 'Proposal for the Expansion of the Panama Canal', ACP April 2006. (ACP)

years. However, final out-turn costs had not been determined at the time of writing and might rise higher than the original estimates. In January 2015 a Conflict Resolution Panel found in favour of the contractor Grupo Unidos por el Canal (GUPC) to the tune of £151 million ($234 million) over the central issues of a billion-dollar dispute, which is yet to be resolved. GUPC is an international consortium composed of the Spanish company Sacyr Vallehermoso, the Italian company Salini Impregilo, the Belgian company Jan de Nul and the Panamanian company Constructora Urbana.

So, what does the completion of the expanded canal achieve? Most significantly 'new-panamax' container ships can carry up to 12,000 TEU containers on each transit, compared to the maximum of 4,500 TEU containers that it is possible for a 'panamax' ship to carry. *Queen Mary 2*, with her length of 345m and beam of 41m, would fit comfortably into the new locks with room to spare. Unfortunately, however, the current thinking at Cunard is that, because of her air draught measured against the quoted clearance beneath the Bridge of the Americas, she is never likely to transit the expanded canal. The ship's keel-to-funnel height is quoted as 236ft 2in and her draft as 32ft 8in. Based on these dimensions *Queen Mary 2*'s air draft would be 203ft 6in, which would prevent passage under the bridge as it has a quoted clearance of 201ft above Pacific high water. However, the Pacific tidal range at the bridge is around 20ft, so the theoretical clearance from the top of the funnel to the underside of the bridge at low water would be 15ft so, perhaps, never say never. Rumours circulating in Panama do suggest that consideration is being given to raising the height of the Bridge of the Americas. Now that would be an interestingly strategic civil engineering project!

It is perhaps a salutary thought that all of the effort put into the Panama Canal expansion project could be undermined if global warming continues to remove ice from the Arctic. If year-round access was opened up to the North West Passage, then this would provide an even shorter route between the eastern seaboard of the USA and Asia. Secondly, in December 2008, reports were being widely circulated in the Russian media that, as relations between Moscow and Nicaragua were getting warmer, President Dmitry Medvedev had become interested in building a canal across Nicaragua. However, in late 2012, statements from Vladimir Putin suggested that, because of global financial issues, Russia was re-evaluating the part it might play in such a project. But then, in June 2013, undeterred by such statements, Daniel Ortega, the President of Nicaragua, announced that a £25 billion rival to the Panama Canal could be built across his country funded entirely by China. Who knows whether either of these scenarios may one day eclipse the Panama Canal expansion project. Only time will tell.

3

CIVIL ENGINEERS IN DEFENCE OF THE REALM

FROM MULBERRY HARBOUR TO FLOATING NUCLEAR SUBMARINE DOCKS

There are three worthy examples of where the skill of the civil engineer has been used to enhance the defence of the realm. Each one of these projects depended for its success on the ability of appropriately designed and constructed concrete to float, perhaps not a property of concrete that immediately comes to mind. The first project is the famous Mulberry Harbour, which was built in north-west France in 1944. The second, taking forward some of the design principles established at Mulberry, is Weston Mill Lake Jetty, a structure that my team designed and built forty-five years after Mulberry at HM Royal Naval Base, Devonport, Plymouth. The third project comprises floating concrete facilities for submarines at HM Naval Base, Clyde.

Mulberry Harbour

As planning for the D-Day Landings of the Second World War was being undertaken, it was realised that the first landings were, of course, to be over open beaches, but also that the invading army would need to be large enough to create a beachhead and then to break out into open country and move inland in considerable force. Such an army would require vast quantities of supplies, vehicles and fighting machines and, to land such equipment, operational harbours would be required. Unfortunately, in 1944, all of the harbours on the northern coast of France were in enemy hands. So in order to test how difficult it might be to capture an operational harbour from the Nazis, an attack on the occupied port of Dieppe was planned for August 1942. The objective of the Dieppe Raid was to seize and hold this major port for a short period, to prove that such an action was possible and to assess the enemy's responses. It was further hoped that useful intelligence would be gathered from captured prisoners.

In the early hours of 19 August 1942, the attacking force – composed of 5,000 Canadians, 1,000 British troops and fifty US rangers

Construction of Phoenix caissons at East India Dock, London. (IWM ART LD 004333)

– departed from the south-coast ports of Portsmouth, Shoreham and Newhaven. Unfortunately, at 3.45 a.m., just after the three elements had joined together in mid-channel, they ran into a German convoy that alerted enemy defences on the shore. One hour later, under cover of a smoke screen laid by HMS *Calpe* and after bombardment of the shore batteries by an accompanying Hunt-class destroyer, the first group of the attacking force, No. 4 Commando, landed at Vasterival Beach. The day turned into an absolute disaster for the Allied force. The RAF failed to lure the Luftwaffe out into open battle but still lost many aircraft. The Royal Navy lost one ship, with others damaged, and over thirty landing craft were sunk. Less than six hours after the start of the attack, Allied commanders were forced to call a retreat. None of the major objectives of the raid were accomplished, and over 3,000 of 6,050 men who made it ashore were killed, wounded or captured. Clearly, if ports were required to land an invasion force, the Dieppe Raid convinced wartime planners that if existing ports on the north coast of France were unavailable then new facilities would have to be built, and built from scratch.

Sir Winston Churchill, from his office in 10 Downing Street, sent a now famous memo to the Chief of Combined Forces, Lord Mountbatten:

Piers for use on beaches. They must float up and down with the tide. The anchor problem must be mastered. Let me have the best solution worked out. Don't argue the matter. The difficulties will argue for themselves.

This short instruction set in motion a mammoth civil engineering design, logistics and fabrication operation that would, in just under two years, change the course of the war and help shape the history of the twentieth century. From 'piers for use on beaches', Mulberry Harbour was born. 'Mulberry' was the code name used for an arti-

ficial harbour that was to be planted on the Normandy beaches in the days following the Allied D-Day invasion of Europe. The only way to create such an artificial harbour was to build it in sections, transport those sections to France and then assemble them together at the chosen location immediately after the first troops had landed on the beaches. Design development started in the summer of 1942 and, at the end of that stage, a detailed model of the finally agreed components was presented to Churchill.

The harbour would essentially comprise a series of jack-up platforms designed to function as pier heads, and these were to be connected to the shore by floating roadways. To protect these unloading facilities from storms, a continuous outer breakwater would complete the harbour. Each of the hundreds of component parts necessary to construct the harbour would be designed to float across the Channel. Final decisions determined that there were to be two Mulberry harbours, not one, and each would be bigger than Dover Harbour. Both were to be located in the Baie de la Seine, between Cherbourg in the west and Le Havre in the east. 'Mulberry A' was to be assembled at St Laurent and was to be operated by the Americans, and 'Mulberry B' was to be assembled at Arromanches, to be operated by British forces.

Military operational criteria demanded that all of the component parts for the harbours would have to be prefabricated within a seven-month construction period. The designs would have to ensure that the parts could be assembled in Normandy in a matter of days. To achieve this staggering programme, multiple standard components would be required. These included the 'Phoenix caisson breakwater unit', which when placed end to end with matching units would build an enclosure to the harbour, providing sheltered water so that ships might safely lie alongside 'jack-up pier head units'. Linked to each of the pier heads would be a 'floating concrete connector unit', from which would spring the 'Whale roadway jetty' making a connection to the land and over which, in large numbers, motor transport and fighting vehicles could drive away from the ships.

Throughout the winter of 1943, and in great secrecy, a project that took priority over all other building construction work was under way at dozens of locations across Britain. Winston Churchill visited many Mulberry Harbour prefabrication sites to encourage the workers at a time when most of them had no idea just what they were building. Suffice to know, he told them that what they were doing would contribute in a major way to the war effort. The strength of the veil of secrecy that surrounded the project was brought home to me when, just before the presentation of one of my lectures on board ship, an elderly lady approached me saying that she was much looking forward to my presentation, as her late father had been in charge of some of the construction processes at 'Mulberry B'. In reply, I suggested that I hoped the details I was about to relate would prove to be accurate, to which she responded by saying that, on the contrary, she knew absolutely nothing about his work, as he had still felt unable, right up to the time of his death, to break the secrecy of Mulberry. After the lecture she approached me again to thank me for providing an insight into how important was the work that her father had been engaged with on those wartime beaches.

Phoenix breakwater units, also known as Phoenix caissons, were built in strategically valuable dry-build facilities, such as dry docks and floating docks, which would otherwise have been used for the urgent repair of ships. The reinforced concrete units would be removed from these facilities as soon as they were able to float, usually when the walls had reached a height of around 7m. Then, tied up alongside any convenient quayside, a further 10m would be added to the walls – such work often being undertaken by small-sized building firms – feeding concrete into the growing structure using wheelbarrows. It was a measure of the absolute priority of Mulberry Harbour that such ship repair facilities, so desperately

needed in wartime, could be tied up for the construction of its component parts. Because cameras were not usually allowed on Mulberry construction sites, official Admiralty artist Sir Muirhead Bone sketched a pair of Phoenix caissons under construction at East India Dock, London. Note that anti-aircraft gun emplacements, a feature of these units, were already in position before the concrete work had been completed. These defensive measures were not just designed for use at the completed harbour but were manned for the journey across the Channel to ward off any aerial attack on the convoy. Another four Phoenix units were built in a gigantic hole that had been excavated alongside the River Thames at Russia Yard, Rotherhithe. At the same time as the hole was excavated, so too was a connecting channel linking it to the river. In order to maintain a dry environment for construction, the end of the channel had been temporarily closed off with a row of steel sheet piles that divers had cut off at bed level once the completed units were ready to be floated out.

Jack-up pier head units had been designed when in use to stand on their own four legs, a technology that has been further developed over recent decades within the offshore oil and gas industry. After being towed into position, the four legs would be lowered to the seabed one at a time, a process that allowed for any irregularities of the seabed. If, for example, one leg happened to be positioned over a hole then that leg would be extended further down until it sat firmly at the bottom of the hole. Once it had been judged that all of the legs had a stable foundation, the floating steel deck was jacked-up to a level position, with its top surface at the height above the water necessary to provide an appropriate mooring face for whatever size of ship was using it. Contained within the steel deck structure were machinery rooms, a galley, mess accommodation and, at one end, sleeping accommodation for an officer, six NCOs and fifteen crew.

Floating concrete connector units were provided to link each pier head to the floating roadway bridge. A connector unit was sized

Jack-up pier head units in operation at Mulberry 'B' Harbour. (IWM A024361)

to allow it to perform as a marshalling area; vehicles that might need to be backed off of a ship had room to turn to face in the right direction for the drive along the narrow roadway to the shore.

Floating Whale roadway jetties consisted of many short steel bridge spans, linked together and supported at each link position. In the jack-up pier head photograph (on the previous page), a steady roadway is apparent for the lorry to drive along, but sea conditions might not always be as calm as that shown. In December 1942, as part of the design and proving process, a prototype MK1 Whale roadway test rig was set up by the Royal Engineers. The roadway was subjected to a moving 20-degree twist to prove that adjoining sections would remain connected under such conditions and to demonstrate that it was still possible to drive a vehicle along it. The roadway structure not only had to be able to cope with whatever sea conditions might persist under operational conditions within the new harbour but also had to deal with conditions encountered on the Channel crossing. Three different types of roadway link support were designed: a steel floating cylinder for use in shallow water; a floating concrete pontoon for use in deeper water; and a steel jack-up platform. The selection of which type of support platform was appropriate in any particular location was based upon the specific beach terrain over which the roadways would run, this having previously been determined by clandestine surveys. The floating concrete support platforms were called 'beetle units'. These precast reinforced concrete pontoons were manufactured on a production line basis. As each unit was completed it was launched using a specially built gantry crane. The small steel jack-up platforms were designed with four spud legs, just like their bigger sister pier head unit. Although jacked up out of the water, once in their operational positions, these support platforms also had to float – a necessity for the journey across the Channel.

Following fabrication, each of the hundreds of the various components of Mulberry were towed or transported to assembly points

Mulberry 'B' Harbour at Arromanches. (IWM BU001020)

on the south coast of England in readiness for the crossing of the Channel. Due to a shortage of mooring places for the Phoenix caissons, many were temporarily sunk in shallow water along the coast to be pumped out and refloated when they were required for the Channel crossing. Long lengths of Whale bridge roadways were assembled together for the tow across the Channel.

At the two Mulberry Harbour sites the outer breakwaters were first constructed by moving individual Phoenix caissons into their required position using tugs; once this had been achieved they were then sunk by opening seacocks, allowing them to fill with water. Later, and as necessary, they were further stabilised by filling them with sand. In addition to caisson units, sunken 'block ships' were also deployed to complete and reinforce lines of breakwater.

Within days, and just as planned, both harbours were in full use. To cope with the rise and fall of the tide, the pier head was jacked up and down on its spud legs to maintain a suitable berthing

Relics of Mulberry 'B' Harbour on the beach at Arromanches.
(Richard Brayden)

face for the ships. At low water the bridge assembly needed to be slightly longer than that required at high water. To accommodate this change, the bridge span nearest to each pier head contained a telescopic section.

Management of the operational piers was complicated by the need to cater for two-way traffic. While supplies were being landed, casualties needed to be moved out.

On 19 June, when the harbours had been fully operational for less than ten days, disaster struck. A summer storm bringing the worst period of sustained severe weather in the English Channel for some forty years caused a major change in operations. Mulberry A was severely damaged, twenty-one of the original thirty-one caissons placed were damaged beyond repair, jetties were damaged and floating roadways displaced. The storm did more damage than the enemy and Mulberry A was never used again. Parts of it were, however, usefully scavenged to repair damage to Mulberry B, which,

luckily, had not suffered so badly in the storm. The Americans quickly reverted to their traditional methods of unloading directly onto the beaches using landing craft and DUKWs (an acronym based on 'D' indicating the model year, 1942; 'U' referring to the body type, utility (amphibious); 'K' for all-wheel drive; and 'W' for dual rear axles), more commonly known as 'Ducks'. Mulberry B Harbour continued to function as designed and contributed greatly to getting that huge army ashore, an army that was to sweep across France and into Germany and to eventual victory.

On 6 June each year, as an integral part of the D-Day commemorative ceremonies centred on the French beach town of Arromanches, the monumental engineering achievement of Mulberry Harbour is celebrated. The scale and complexity of designing a harbour system built in one country to be used in another and capable of landing hundreds of thousands of troops and all of their equipment will probably never be repeated. Many of the once floating components

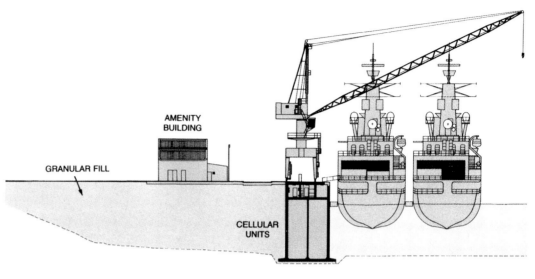

Logo of PSA Defence Works Navy.   General cross section through Weston Mill Lake Jetty. (DWN)

of Mulberry can still be seen on the beach at Arromanches today, standing testament to those construction workers who literally cast their place in history back in the winter of 1943.

Weston Mill Lake Jetty

As civil engineering director of Defence Works Navy, I, along with my team, was responsible for the project management and delivery of Weston Mill Lake Jetty at HM Naval Base Devonport, Plymouth. The structural form of the jetty, somewhat unusual for its time, might be seen as a development of some of the design concepts used at Mulberry Harbour. Weston Mill Lake is, in fact, a wide inlet off the tidal River Tamar at Plymouth, at a point where the river is known locally as the 'Hamoaze'. Our brief was to design and build a major new facility that would permit up to six Type 22 frigates at a time to undergo both self-help and assisted maintenance regimes. The design was also to afford the opportunity to reclaim from the muddy river some 32 acres of land to supplement the area available

for future naval-base development. The team first undertook the initial appraisal of the Ministry of Defence project brief to determine the functional requirements and special needs of the jetty. The structural form of the facility was selected after detailed technical and financial evaluation of no fewer than six different forms of construction using both structural steelwork and concrete design solutions. A solid jetty scheme was taken forward by the design team in an intensive forty-week period in 1985. Contract documents were drawn up and my office sought tenders from eleven international construction companies. A contract was awarded to Costain Civil Engineering in the autumn of 1986.

The most important elements of the chosen structural solution were precast reinforced concrete cellular units (caissons). The design called for twenty-nine units (including four special link units), each 10m wide, 13m high and 19m long. Placed end-to-end, they were to create a berthing face nearly 500m long. A reinforced concrete services subway was constructed on top of the caissons to tie them together. The roof of the services subway formed the front part of the jetty deck area. The remaining deck area would be the top of a body

Dredger *Bilberg 1*. (Author's collection)

Concrete caisson casting yard at maximum production with runway three full of caissons awaiting their turn to be launched. (Author's collection)

of imported granular fill material placed behind the berthing face. To complete the project, a group of amenity and service buildings and two 10-ton luffing cranes were to be added to the deck area.

The jetty was to be constructed in a sequential operation, but the project started with an advanced works dredging contract. A wide and level area was prepared along the floor of the lake, over which the caissons would be floated before being sunk into position. For this part of the project we used the dredger *Bilberg 1*, then the largest dredger of its type in the world.

The reinforced concrete caissons were to be constructed at a specially prepared casting yard on the riverbank at the north side of the lake. As each was completed, it would be floated into position, in a similar way that the Phoenix caissons of Mulberry Harbour were

handled some forty-five years earlier. Unlike the Mulberry units, where the dock in which they were built was flooded in order to float them, our caissons, each of a mass equivalent to a four-storey apartment building, would, on completion, be launched down a slipway.

To achieve an efficient sequential construction programme, the prefabrication yard was set up like a factory, with two tower cranes on site to handle all materials, which included the 2,000 tons of concrete that went into each unit. Two identical side-by-side production lines, known as 'Runway 1' and 'Runway 2', were set up. At the head of each runway, furthest from the river, was the 'base construction bay'. Once a caisson base had been cast, it was then pushed by a hydraulic ram onto a 'base storage bay', allowing fol-

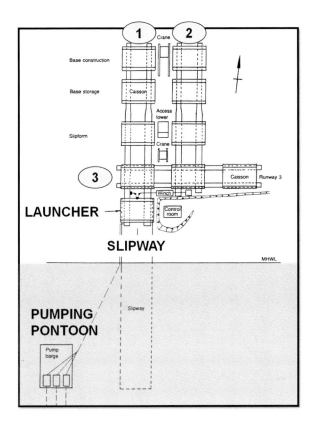

Plan of the caisson construction 'factory'. (DWN)

A caisson sitting on the launcher awaiting its journey down the slipway. (Author's collection; DWN)

low-on bases to be cast in the first bay. As a wall construction bay ('slipform bay') became available, a caisson base was moved onto it from the storage bay. Work on the reinforced concrete walls of the units was then undertaken using a continuous non-stop concrete pour. Once that process had been completed a finished caisson was moved onto 'Runway 3', which linked the other two runways and provided an additional parking bay for a completed caisson to await its turn to be launched down the slipway into the river. The two production lines were run seven days out of phase with each other, thus establishing a continuous production regime with time allowed for the concrete to cure sufficiently so that units could be moved between the construction bays without damage. Concrete curing is the chemical process that starts once water has

been added to cement and ends when the concrete has gained its designed strength.

When ready, each unit was moved in turn onto a 'launcher' for its journey down the slipway into the river. Just offshore from the launching slipway, a pontoon barge loaded with powerful pumps fed high-pressure water to the underside of the launcher, which floated down the slipway on a cushion of water carrying the caisson on its back. Once these huge concrete caissons were afloat, any similarity with the Mulberry Harbour breakwater units then ended. Remember that at Mulberry, the Phoenix caissons were placed together with a loose fit and just sat on the seabed at whatever level it happened to be. At Weston Mill Lake, the floating concrete units were carefully manoeuvred into position using tugs, with each unit being precisely aligned to its neighbour. Two vertical keyways formed during caisson casting were aligned with corresponding keyways on the end of the adjacent caisson. The units were then slowly sunk into position by pumping water into them, with small adjustments of position being continuously made until the unit grounded onto a bed of crushed stone that had been placed on the river bed, the level of which had been precisely prepared by the dredger *Bilberg 1*. Once in its final position, mass-concrete shear keys were formed underwater in the

Units in position on the lake bed at various stages of completion. (Author's collection)

enclosed keyway slots. In order to prevent segregation of the concrete it was placed using a tremie tube, which is arranged so that the underwater foot of the tube is always kept below the level of the concrete being placed and the tube is always kept full of concrete, with its upper surface maintained above water level. In this way the concrete in the pipe is always under a higher pressure than the water outside it and is therefore unlikely to be diluted by water. The projecting toe, formed as part of the caisson base, was then protected by a rubble mound that was itself keyed into the riverbed.

One by one, the twenty-nine units were placed in line starting furthest from the River Tamar. As soon as the first few units had been placed, then another 'production line' sequence of operations began. The most recently placed units were accurately positioned and filled with seawater to lock them to the riverbed. Those placed just before had already been filled with sand, which had displaced the temporary water filling whilst the tops of still earlier-placed units were being prepared to start the construction of the services

subways and jetty deck. On the top of those placed before that, the services subway was actually under construction. Landfill was progressed behind the earliest-placed caissons when completed with their services subway on top. All fill material was dredged from Plymouth Sound by suction dredger and then hydraulically deposited behind the cellular units. The sequential construction process continued until the full jetty length had been formed.

The three-chamber services subway was a continuous in situ concrete structure that provided a positive tie to the independent caisson units. The top of this subway formed the quay surface of the new wharf. Supplied from dockyard systems through the subways to ships alongside the jetty are steam, diesel, freshwater, telephone and high-voltage electrical services. Produced in the jetty service buildings and delivered through the subway are saltwater, chilled water, compressed air and three-phase electrical services. A ship to shore foul-water link is provided back through the subway feeding into dockyard systems.

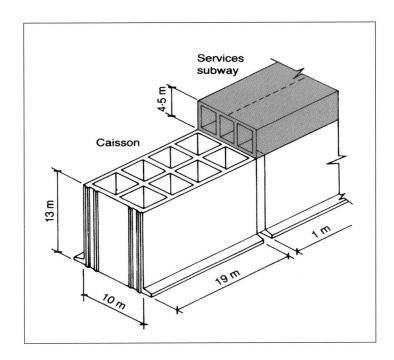

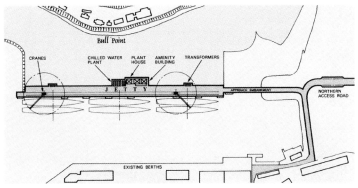

Above: Plan of the completed jetty and landfill area. (DWN)

Left: Relationship between the caisson units and the in situ services subway. (DWN)

Work on the amenity and service buildings was progressed as soon as possible after the fill material had been consolidated. To link the jetty to the existing dockyard facilities, a 180m long approach embankment, faced with rock armouring, was constructed. The embankment provided access to the jetty for vehicles and made provision for the connection of the mechanical and electrical services routed through the subways. To complete the project in accordance with the Ministry of Defence brief, we took delivery of, and erected, two 10-ton level luffing cranes. At a ceremony on 14 April 1989, after some 58,000 tons of concrete had been made to float in the spirit of Mulberry Harbour, the facility valued at £16 million at then current prices was formally opened by Admiral Sir Julian Oswald.

Facilities in Scotland for Submarines

Another set of projects, providing facilities in Scotland for the Astute and other classes of submarine, once again depended on the ability of appropriately designed concrete to float.

A satellite view of part of the naval base at Faslane on the Gare Loch in the Firth of Clyde shows Valiant Jetty, the pier at the centre of the photograph on page 56. This 200m long by 28m wide by 10m deep reinforced concrete structure was designed to provide berths for the Astute class of submarine, a nuclear-powered attack submarine. (When HMS *Astute*, the class leader, ran aground near Skye Bridge in Scotland in October 2010 it generated headlines across the media. One detail that pictures taken at the time clearly demonstrated is that a modern nuclear-powered vessel of this type is in fact a steamship. Often overlooked in the twenty-first century is the fact that this means of propulsion has survived from the early days of the industrial revolution; it is merely the method of boiling the water to make the steam that has changed.)

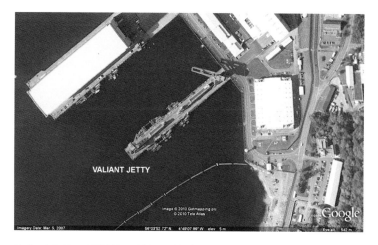

Valiant Jetty at HM Naval Base Clyde. (Google Earth)

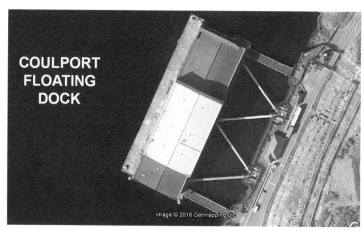

Floating dock at HM Naval Base Clyde. (Google Earth)

Valiant Jetty, unlike most jetties, is not supported on legs sitting on the seabed but is a massive floating concrete structure built at a cost of £150 million at Inchgreen Dry Dock, Greenock, from where it was moved by five tugs to Faslane. It is now permanently held in position by four retaining piles (each the height of Nelson's Column in Trafalgar Square, London), which allow the jetty to rise and fall with the 3m tidal range of the Gare Loch. Valiant Jetty clearly demonstrates the principal advantage of a floating jetty (particularly where sensitive work is being undertaken on ships alongside) of being able to maintain a fixed height of jetty deck above the water no matter what the state of the tide.

On the other side of the range of hills separating the Gare Loch from Loch Long is another associated Ministry of Defence facility. Just as some of the Mulberry Phoenix caissons were built in holes in the ground next to rivers back in 1943, so this structure was built in a large hole in the ground excavated into an area of reclaimed land at Hunterston near the mouth of the Firth of Clyde. Once this massive structure had been completed, the earth bank separating the construction site from the Clyde and maintaining a dry construction environment within the hole was dredged away, allowing water from the river to flood the work site and float the structure.

After being towed 32km to the operational site at Coulport, the floating structure was permanently held to the shore by four tubular-steel mooring booms. The mooring booms were fixed to spherical bearings on the structure and to articulating hinges on the shore, allowing the whole covered dock structure to move up and down with the tide. This movement allows submarines moored inside the building to be serviced no matter what the state of the tide, as the freeboard or height of the dockside above water level remains constant at all times. The facility is accessed by two articulating road bridges, one at each end.

The Future for Floating Concrete

The science of 'very large floating structures' (VLFS) is now a branch of civil engineering in its own right. VLFS can be constructed to create floating airports, bridges, breakwaters, piers and docks, offshore storage and production facilities (for oil and gas), wind and solar-power plants and even individual dwelling houses or larger floating communities where people might live – a very long journey from the floating structures of Mulberry Harbour.

4

EUROPE'S LARGEST SHIP CANAL

ONCE THE EIDERKANAL NOW THE KIEL CANAL

The Kiel Canal has a fascinating history. It started out life as the Danish Eiderkanal, then became the German Kaiser Wilhelm Canal and later, as we know it today, the Kiel Canal. The waterway is located in Germany, linking the Baltic Sea to the North Sea. The Kiel Canal is not only the largest canal in Europe but is also quite unique for a ship canal as it is surrounded by lush fields, with cattle grazing right up to the water's edge and, as ocean-going ships transit through villages, residents wave from the bottoms of their gardens or from the balconies of their houses.

Before the building of an inland navigation across the Jutland Peninsular, the route taken by ships travelling from the English Channel to the Baltic Sea was across the North Sea, through the storm-prone narrows of the Skagerrak and the Kattegat to the north of Denmark, with entry to the Baltic being through the narrow seaway between Denmark and Sweden. In the days of sailing ships, this was a particularly difficult passage. Historic maps record the position of over 6,000 wrecks occurring between 1858 and 1885, that's an average annual loss of over 200 ships, or four every week. 'Wind Jammer' bay in Skagen was the terror of sailors, being one of the most notorious coastal waters on earth. Once built, the waterway provided a safe route across Jutland, reducing a ship's passage between the seas by at least 250 nautical miles. Today, its operators claim that the Nord-Ostsee-Kanal (or NOK, as it is known in Germany) is the most heavily used artificial waterway in the world.

Man-made waterways were built on Jutland long before the Eiderkanal came into existence. The very first was built in the fourteenth century to transport salt from Lauenburg to Lübeck. This waterway, effectively joining the Rivers Elbe and Trave, was further developed between 1895 and 1900 to provide a through route between the Baltic and the North Sea for ships smaller than the Kiel Canal could accommodate. A direct waterway from the city of Kiel on the Baltic Sea across the peninsular to Tonning on the North Sea, via the River Eider, was first proposed in 1571 by Duke

Mooring posts marking a siding on the Kiel Canal.

Adolf of Holstein-Gottorp, but it would be another 200 years before construction of such a waterway would start.

In 1777, a project was begun to improve navigational facilities on the River Eider and then to extend the navigation eastwards to Kiel by the construction of a new length of canal. The improved river navigation from Tonning at the North Sea mouth of the River Eider to Rendsburg was 132km long and the new canal section from Rendsburg to Kiel-Holtenau was just 43km long. The completed navigation from sea to sea was to be known as the Eiderkanal. The designed line of the new channel between Rendsberg and Kiel-Holtenau was provided with five locks to raise and lower the waterway over high ground and incorporated and linked up with four existing lakes along its route. Some 4,600 construction workers were employed on the project with, sadly, more than half of that number dying from diseases contracted during the works. It has been estimated that, to construct the 29m-wide by 3m-deep waterway, those 'navvies' moved by hand nearly 100 million cubic metres of material.

Seven years after the start of the works, the Eiderkanal was opened by the Danish king, Christian VII. Thus was established the first inland navigation between the North Sea and the Baltic Sea. To understand why a Danish king opened the Eiderkanal, the map above may be of some help, as it shows the boundary between Denmark and Germany prior to the German–Danish War of 1864, clearly illustrating that the canal was built entirely on Danish soil.

For over 100 years, the Eiderkanal provided a safe route for sailing vessels, rather than that through those stormy waters to the north of Denmark. In 1872, reckoned to be the peak year of its operation, over 5,000 vessels transited the waterway. Passage from sea to sea took from three to four days, with vessels towed along the canal by horse. The importance of the canal as a safe route for sailing ships diminished with the advent of steam power. Tugs that were used to berth ships at the many wharves along the

The boundary between Germany and Denmark prior to the German Danish War of 1864.

waterway were later utilised to tow sailing ships the entire length of the canal.

Eventually steamships took over from sail – the drawing on page 60 depicts an example of the largest steamship able to transit the waterway. Built in 1886, the SS *Kanal* was sized to just fit the locks of the Eiderkanal, just as 'panamax' ships were built to just fit the Panama Canal locks. By the late nineteenth century, it was clear that the dimensions of the locks, just 35m long, 7.8m wide and with a depth of 4m, were simply not big enough to accommodate the size of vessels then being built and, perhaps of even more importance, it became apparent that the canal could not satisfy the new political aspirations of a Europe in turmoil.

Throughout the life of the Eiderkanal, the Jutland Peninsular had seen wars come and go. In 1848, an independence movement in Schleswig-Holstein demanded secession from Denmark. Forcefully put down by the Danish government in the First Schleswig War, the Schleswig-Holstein forces had to abandon their independence plans. In 1863, Danish politicians who were against international agreements existing at that time planned to incorporate the dukedom of Schleswig into the Kingdom of Denmark. The following year, Prussia and Austria jointly attacked Denmark in order to push it out of the northern provinces of Schleswig and Holstein. The German–Danish War of 1864 was the first in a series of three German 'wars of unification' (the others following in 1866 and 1870–71) and was an absolute disaster for Denmark, which lost the duchies of Schleswig to the Prussians and Holstein to the Austrians. A victory column in Berlin, designed by Heinrich Strack to commemorate the Prussian victory over Denmark, was inaugurated in 1873, but by that time the joint victors of the Danish conflict had already gone to war with each other over those same territories. Prussia defeated Austria in 1866, going on to defeat France in the Franco-Prussian War of 1870–71.

At the beginning of the Franco-Prussian War, the North German Federal Navy became the Imperial German Navy – the Kaiser's Navy. A new naval base at Wilhelmshaven on the North Sea had just been completed and at Kiel on the Baltic Sea there already existed a state naval harbour. The Kaiser desired to maintain powerful fleets in both the North Sea and the Baltic Sea and he knew from experience just how easily the passage through the narrow channel between Denmark and Sweden could be blocked in time of war. The two fleets would either have to be able to operate independently of one another or a safe passage would have to be provided so that they could reinforce one another in time of conflict. Thus devel-

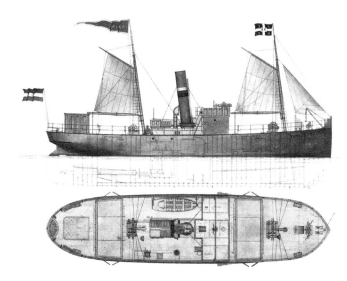

General arrangement drawing of the steamship *Kanal*. (Dr Jürgen Rohweder)

oped a strategic argument for a canal that would be large enough to link the two naval bases. The Eiderkanal was much too small for that role; therefore, an enlargement and partial re-routing of the waterway were proposed. Because the frontier between Denmark and Germany had been redrawn at the end of the German–Danish War (it having moved to the north) the re-engineered Eiderkanal, now to be known as the Kaiser Wilhelm Kanal, would sit entirely upon German soil.

So it was, then, that on 3 June 1887, twenty-three years after the end of the German–Danish War, Kaiser Wilhelm I of Germany laid at Holtenau near Kiel the foundation stone of his North Sea–Baltic Sea canal. From Kiel westwards, it closely followed the alignment of the eastern end of the old Eiderkanal. However, in order to improve the navigation prospects for the larger ships that would be using the waterway, civil engineers took the opportunity to build five major cut-off loops that were located at Levensau, Sehestedt, Budelsdorf, Rendsburg and Schulp. Pre-construction planning of the new sections of waterway was complicated by the requirement that the existing navigation had to remain operational until the improved waterway had been completed. Beyond the village of Schulp, the

Grunental Railway Bridge – a wood engraving by Fritz Stoltenberg.
(Dr Jürgen Rohweder)

re-engineered waterway followed an entirely new south-westerly route towards the North Sea at Brunsbüttel on the Elbe estuary.

The entire length between Kiel-Holtenau and Brunsbüttel was designed to be at the same level throughout, with locks only being provided at each end in order to cope with the varying tide levels of the two seas.

The locks were constructed in excavated pits, with spoil being carried away by temporary railway lines. At Brunsbüttel, because of an exceptionally high groundwater table, continuous day and night pumping was necessary in order to allow building work to proceed. These particular locks were not only designed to cope with tidal variations but were also designed to prevent North Sea surges from entering the canal and flooding the low-lying countryside along the line of the waterway. Some 9,000 workers laboured for eight years to re-engineer the canal. Thirty 'dry excavators', each consisting of twenty or so curved steel buckets mounted on a continuous chain and driven by a steam engine, were deployed on the side slopes of the channel. As the buckets reached the top of their travel, they inverted, tipping their load into a slowly moving train of railway wagons. The sequence of construction was first to excavate the

upper portion of the canal and then to move the whole operation downslope by creating a series of ledges cut into the side slope, upon which railway track would have been laid, and the excavators located. What a contrast this mechanised construction was when compared to that used to build the original Eiderkanal, where pickaxes and wheelbarrows were the order of the day!

However, if an early wood engraving by artist Fritz Stoltenberg is an accurate representation of the work being undertaken near the railway bridge at Grunental then, despite the use of heavy construction equipment, it appears that pickaxe and shovel work was still being employed in some areas (bottom left-hand corner of picture). Some 800 million bricks were needed for the work and dedicated brickworks were set up to produce 40,000 bricks every day. Granite blocks, required to provide 'rip-rap' protection to the exposed banks of the canal, were quarried locally. Final navigational depth was achieved using floating bucket dredgers, with spoil being taken away by barge and dumped at sea. When completed in 1895 the re-engineered waterway was trapezoidal in shape measuring 67m wide at the surface and with a depth of 9m. The locks at each end of the canal were built in pairs, each being 25m wide. The lock walls at Brunsbüttel were built higher than those at Holtenau to allow for tidal surges of the North Sea. Commercial ships of the day could easily pass one another in most lengths of the canal but not so the large ships of war; therefore, passing places were constructed at appropriate locations.

A two-day ceremony to mark the opening of the re-engineered waterway started on 20 June 1895 when Kaiser Wilhelm II entered the canal at Brunsbüttel aboard the German royal yacht *Hohenzollen*, then the largest royal yacht ever built. Completed in 1893 by the Vulcan Shipbuilding Company at Stettin, she had a length of 117m, a beam of 14m and a draught of 7m. Her 'Thames tonnage' was given as 3,773 and she had a top speed 21.5 knots. All along the route, thousands of people lined the canal banks to wave

German submarines at Kiel Naval Base, 1914. (Author's collection)

as the Kaiser sailed past. Hundreds more crowded the lock-side for his arrival at Kiel-Holtenau where, on the second day, before many of the crowned heads of Europe and to the strains of massed military bands, Kaiser Wilhelm laid the final ceremonial stone of the reconstructed waterway, which he formally named the Kaiser Wilhelm Kanal, the name by which the Kiel Canal continued to be known until 1948. British director Bert Acres filmed the opening of the canal and surviving footage of this early film is preserved at the Science Museum in London. Ships of many nations attended the grand opening ceremony including no fewer than nine ships of the Royal Navy (according to a list published in *The Times* of 21 June 1895): HMS *Royal Sovereign*; HMS *Repulse*; HMS *Resolution*; HMS *Empress of India*; HMS *Bellona*; HMS *Speedy*; HMS *Blenheim*; HMS *Osborne*; and HMS *Enchantress*.

Designed to accommodate the largest warships then in existence, the completed canal initially satisfied the strategic aspirations of the German emperor by linking those principal naval bases at Kiel and Wilhelmshaven. However, barely ten years after the opening of the re-engineered canal, the Royal Navy in 1906 launched the first British dreadnought, a new type of battleship. In an instant, this single act rendered the strategic value of the Kaiser Wilhelm Kanal obsolete. All navies, including that of Germany, were forced to follow the lead of the Royal Navy and build ships of this type and size; unfortunately for Imperial Germany, ships of this size were too large to transit the new canal.

Undeterred, however, the German government immediately put in hand a programme of deepening and widening of the canal. So as to interfere as little as possible with the existing navigation, much of the widening work was undertaken on one side of the canal only. Most of the deepening was undertaken by floating bucket dredgers, with the spoil being removed by barge or by railway and much of it being dumped at sea. The amount of spoil removed in this enlargement operation was said to have exceeded 130 million cubic yards, more than from the entire excavation of the earlier Eiderkanal. Completed in 1914, the enlarged trapezoidal cross section measured 102.5m wide at the surface and 11m deep. At several locations, the canal was further widened to allow very large ships to pass one another. The lengths of these passing places varied between 600m and 1,000m. The enlargement project included the provision of an additional pair of larger locks at each end of the canal, which measured 310m in length and 40m in width. (The new locks were of a slightly larger size, in plan, than those at the same time being built on the Panama Canal.) The 1895 locks remained in operation for the use of smaller ships, so there were now four locks available for use at each end of the waterway. The canal was once again able to transit the largest warships in the world, but within months of the completion of the enlargement project the First World War broke out.

For most of the 1914–18 war, the canal served the Imperial German Navy well, providing a link between their North Sea and Baltic Sea fleets. At the beginning of hostilities fifteen German submarines were photographed together at Kiel. In the picture opposite, the second from left in the front row is the *U-20*, which on 7 May 1915, under the command of Kapitänleutnant Walther Schwieger, sunk the Cunard liner RMS *Lusitania* off the south-east coast of Ireland. It is well documented that this act contributed greatly to Germany's defeat in the First World War, as the tragedy helped to bring the USA into the conflict. The U-boat on the right, again in the front row, is *U-21*, which, under the command of Kapitänleutnant Otto Hersing, went on to sink the British battleships HMS *Triumph* and HMS *Majestic*.

After the war, because of its military and commercial importance, the canal was 'internationalised' by the Treaty of Versailles of 1919; the day-to-day administration of the waterway was left with the German authorities. By restricting the size of the German Navy, the treaty dealt a severe blow to the economy of Kiel as a major part

Rolling lock gates at Kiel Holtenau. Note the bomb-proof housing into which the gates are moving.

of the city's 250,000 population depended heavily upon work in the naval shipyards. It was not until the 1930s, with the rise of the Third Reich and expansion of the German Navy, that prosperity once again began to return to the area. In August 1936, Kiel was the venue for the sailing events at the Olympic Games, the race courses for smaller boats being set in the inner fjord. To avoid interference by commercial traffic during the events, the canal was closed for the duration of the games. Adolf Hitler used this period of closure to remove the international status of the waterway, announcing upon its reopening which countries would be allowed to use it. During the inter-war period of German naval expansion, the Kiel shipyards built vessels of all classes, from battleships to fast patrol boats and submarines. The largest battleship built at Kiel was the *Gneisenau* (32,000 tons), which was launched at the Deutsche Werft yard on 8 December 1936, entering into service in May 1938.

During the Second World War the existence of the canal and the strategically important shipyards at Kiel rendered the city an important target for the Allies. One of the most notable of the Allied operations was the raid carried out on the night of 1–2 July 1942 on the *Scharnhorst*, which was being repaired in a floating dock. Three direct hits were observed on the ship and others on the floating dock. It was reported that burning oil was visible 129km away. On 12–13 May 1944, Bomber Command's No. 692 Squadron targeted the Kiel Canal, with route and target marking having been carried out by No. 139 Squadron. Other Pathfinder Mosquitoes attacked the locks at Brunsbüttel, while intruders from No. 100 Group shot up gun positions along the canal. The mine-laying operation, executed by No. 692 Squadron in bright moonlight and from very low level, was a complete success; the Kiel Canal was completely closed to traffic for seven days, by which time sixty-three ships had been held up. On the night of 9–10 April 1945 another major raid, composed of 591 Allied bombers had amongst its targets a huge concrete bunker built in the autumn of 1943 to shelter U-boats.

Despite being hit several times during numerous earlier air raids, it had suffered little damage, but on this occasion it was severely damaged and *U-4708*, which was sheltering inside, was destroyed. Throughout 1945 there were many raids on the naval base and many ships were sunk, including the pocket battleship *Admiral Scheer*. By the end of the war, about 80 per cent of Kiel had been destroyed. At General Eisenhower's headquarters in Reims, France, on 7 May 1945, General Alfred Jodl, Chief of the Operations Staff in the German High Command, signed the document of unconditional German surrender. Shortly afterwards, the canal was reopened to all traffic and its peacetime economic success grew with the wider German economy.

Between 1962 and 1966, at specific locations where reconstruction had been necessary due to erosion of the canal banks, the opportunity was taken to further widen the canal and to reduce the angle of the side slopes, thus lessening the risk of further erosion. Over 40,000 vessels, excluding small craft, have been recorded passing through the Kiel Canal annually, making it the busiest ship canal in the world. Although the plan size of the locks on the Kiel Canal is larger than that of the locks on the Panama Canal, 'panamax' ships are unable to transit the waterway due to the relatively low height of the bridges, which are set 42m above the water.

In addition to the four-lock canal entrance from the North Sea at Brunsbüttel and the four-lock entrance from the Kiel Fiord at Holtenau, it is possible for smaller vessels to join the Kiel Canal about halfway along its length at the village of Oldenbüttel, where the short Giesel Canal links to a still navigable part of the Eiderkanal. The 1895 locks, built in pairs at each end of the waterway, were built with, and still retain, lock gates of the mitre door type, common on British canals and the same as those used on the Panama Canal. The use of mitre gates necessitates the lock chamber being built longer than that required for a ship since, when open, the gates themselves take up space along the lock wall. This is one of the reasons why the

Canal control building located between the 1914 locks at Brunsbüttel.

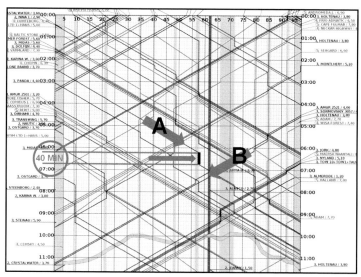

Electronic ship management chart. (Author; NOK Authority)

new jumbo locks of the Panama Canal expansion project, which was completed in 2016, have been provided with rolling lock gates instead of mitre gates.

The rolling gates will slide into recesses at right angles to the lock side, allowing ships to pass. These recesses have another use: when closed off at their outer end they function as a small dry dock, within which, when pumped dry, maintenance of the gate may be undertaken without the need to remove it from the lock chamber. The Kiel Canal was, however, first to adopt this technology. The 1914 locks were provided with rolling lock gates that slide out from recesses to close the lock and slide back home again to permit a ship to pass. Each gate is designed to resist pressure from either side. At low tide, the gates have to hold back the pressure of water from within the canal and at high water, plus an occasional storm surge, they have to resist the higher pressure of water trying to push into the canal from the sea. These open-structured, electrically operated gates weigh up to 1,300 tons each and when fully retracted into their housing provide no obstruction to shipping movements. Some of the housings still retain the heavy reinforced concrete bombproof covers, installed during the Second World War to protect the gate in an air raid.

Located on the central island separating the 1914 locks at Brunsbüttel is the control building for the entire canal operation. All vessels navigating the canal are allocated to one of six 'traffic groups' determined by their size and the type of cargo being carried. In the control building the movement of all ships is managed on a sophisticated time/distance screen that shows electronically, using satellite positioning, the location of every vessel in transit in real time. Alongside the name of each ship, the traffic group to which it has been allocated is shown. Time is measured down the screen, the left-hand side of which represents Brunsbüttel and the right-hand side Kiel. The distance between these two places is measured across the chart: 0km at Brunsbüttel, 97km at Holtenau. The darker-shaded vertical stripes represent the ten passing places where ships of appropriate traffic group may safely pass one another. The canal varies in width throughout its length and depending on its width at a given location the stretch of canal at that location is allocated one of three 'passage numbers' – 6, 7 or 8. The computer

One of the ten passing places, or sidings, provided with lines of bollards to which halted ships may make fast to allow a convoy moving in the opposite direction to pass.

then determines whether two ships approaching one another may safely pass in a given length. It does this by adding together the traffic group numbers allocated to each of the approaching vessels. If the resulting total is greater than the passage number for the length of canal in which they would theoretically meet, then the resulting meeting would be unauthorised. In the example on page 66, ship A is travelling from Brunsbüttel to Kiel and ship B from Kiel to Brunsbüttel. The computer has calculated that their theoretical meeting would be in a length of canal not wide enough for them to pass one another safely. Ship A therefore needs to be stopped at the passing point ahead of this theoretical meeting point, represented in the diagram by the vertical line on the ship's track, indicating that a stop of forty minutes has been computed. Such a decision is then conveyed to ships out on the canal using a system of traffic lights. Each of the ten passing places, or sidings, is provided with lines of bollards to which halted ships may make fast. Today, however, with the accurate positioning ability of vessels using bow and stern thrusters, it is more often the case that a ship might just simply heave-to, while another ship, or convoy of ships, passes in the opposite direction.

All ships, regardless of size, have to take on a pilot for a transit; in fact, two pilots are used. Halfway along the canal is a pilot station where the pilot who has brought the ship through the first half of the canal hands over to a colleague who takes the ship the remainder of the journey. The average passage time for a ship, including waiting at the locks and, where necessary, in the passing places, is between eight and nine hours. Some navigation instructions presented as simple pictograms include speed limit signs; 'watch your wash' – an instruction to avoid making waves that might damage sensitive banks or upset moored craft; 'anchoring is prohibited' – signs that are likely to be sited where cables or other services are buried in the bed of the canal; and 'approach no closer' than the distance in metres to the bank indicated by an arrow. A common traffic sign

A fully loaded ferry at Breiholz waiting to cross the waterway.

seen along the canal is the one that gives advanced warning of a ferry crossing ahead.

There are no fewer than twelve busy ferry crossings along the canal, some having a single ferry plying to and fro, while on busy routes pairs of ferries cross midstream. It is quite entertaining to observe the ferries nipping across between transiting ships moving in convoy. Pairs of red and white diamond marker boards indicate the position of the navigation channel beneath bridge structures. There are ten fixed bridges that cross the canal, some of which are very old, historic structures and others state-of-the-art modern examples of civil engineering. It is the height of these bridges above the waterway that principally restricts the size of ships able to use the canal.

A closely spaced pair of modern, steel, box-girder bridges are the first encountered when entering the canal from the Baltic. Known as the Holtenau Road Bridges, they carry National Expressway 503 across the canal. The later of the two bridges was completed in 1996 and was built next to the Olympic Bridge, which was completed in 1972. These bridges replaced an earlier crossing, the 1912 Prinze Heinrich Bridge, which was removed in 1992.

Next to come into view on a westward transit is another closely spaced pair of bridges at Levensau. The newest one of the two is a

Levensau bridges.

road bridge carrying National Expressway 76, which was opened in 1984 and shares space with the modified Levensau High Bridge. Emperor Wilhelm II laid the foundation stone of this rail, road and pedestrian crossing on 24 June 1893 and opened the bridge just eighteen months later, a full year before the canal beneath had been completed. The main structure is a double 163m wrought-iron arch truss, which at the time of its construction was the longest arch bridge in Europe and the wider German empire. From the outset the bridge had been designed as a grand structure. At each end, an impressive pair of 70m-high masonry towers were linked with an arched portal. Unfortunately, the space between the towers was limited and when a train occupied the portal the road had to be closed. Out on the bridge span, a guardrail between the road and the railway had been installed in order to permit simultaneous use. However, this left the road with only one single, 4.2m-wide carriageway alongside a 1m-wide pedestrian walkway. Faced with increasing road traffic and severe constraints of usage, a decision was made in 1954 to rebuild the bridge. First the grand masonry portals were taken down stone by stone, followed by the removal of the entire bridge deck, leaving just the the original double wrought-iron arch in place, upon which a brand-new bridge deck was constructed. Throughout a well-managed project, canal traffic continued to operate, suffering very little disruption. The new deck was wider than the original, being designed to carry a single-track railway and a two-way single carriageway road between the arch girders, with the pedestrian walkways now cantilevered on to the outside of the main arches. Today, the railway line is a main commuter route carrying modern double-deck trains in the sky above cruise ships. Still a stunning engineering spectacle, the bridge has one other claim to fame. From mid November each year, more than 5,000 noctule (*Abendsegler*) bats, one of the largest bats of the Schleswig-Holstein area, arrive to overwinter within the bridge structure, departing the following March. Naturalists believe this to be one of the biggest community hibernations of its kind in Europe.

Continuing a westward transit, the next bridge encountered is a double-plate girder steel bridge providing a crossing for Autobahn 7. This crossing also bridges the original Eiderkanal just to the north of the present waterway.

A little further along the Kiel Canal is the famous Rendsburg Railway Bridge. This railway bridge is also a road 'transporter bridge', one of only eight surviving in the world, sixteen having originally been built. The other surviving bridges are located near Bilbao, Spain; Hamburg, Germany; Rochefort, France; at Buenos Aires, Argentina; and at Middlesbrough, Newport and Warrington in the United Kingdom. Designed by Civil engineer Friedrich Voss, the Rendsburg Bridge was constructed between 1911 and 1913 to carry the Neumünster–Flensburg railway over the canal from the relatively flat land on either side. Together with its associated viaducts, Rendsburg High Bridge is the most prominent structure in the town and, at 2,486m (8,156ft) long, it is the longest railway bridge in Europe. On its northern side, the bridge connects to the Rendsburg Railway Loop. This 360-degree oval of track allows trains to gain height to cross the high girders after they have departed from Rendsburg Station, which is located very close to the canal. The railway bridge with its underslung 'road ferry' is much celebrated in Germany and has more than once appeared on the country's postage stamps. At the base of the bridge is the famous Ships Welcome Point restaurant, where an announcer gives details of passing ships to diners, usually playing the national anthem or popular music from the country to which the transiting ship belongs.

Just to the west of the railway bridge there are two other crossings of the canal, not readily seen from the deck of a ship as they are both in tunnels excavated below the waterway. A dual carriageway road tunnel completed in 1961 is celebrated on yet another postage stamp, this time commemorating the 75th anniversary of the opening of the canal. The other crossing is described in local tourist guidebooks as one of the longest tunnels for pedestrians and

Rendsburg Railway and 'Transporter' (Road) Bridge.

Adonia passing beneath Grunental High Bridge.

cyclists in the world. Access to this tunnel is via deep escalators or by lifts. Open twenty-four hours a day, the escalators start moving automatically as soon as anyone approaches. Together these tunnels replaced the Rendsburg Swing Bridge of 1913. The present channel through Rendsburg is, in fact, one of those major cut-off channels constructed when the canal was re-engineered between 1887 and 1895. The original and much smaller-scale Eiderkanal lies just to the north and west.

Further west, the next bridge is a very neat and simple steel Warren type through-lattice truss bridge, the Grunental High Bridge, which carries both road and rail across the canal. Constructed in 1986, this modern structure replaced another ornate bridge of the same name, which was completed in 1892. Once the replacement bridge was operational, the old one was dismantled, the double arch being lifted out in one piece by two huge floating cranes that completely blocked the canal. The arch was then moved to a prepared flat area alongside the canal where it was broken up. The original abutments that once supported the arch are still in position but are now difficult to see from the canal due to extensive vegetation growth. There is not a lot of room to spare under any of these bridges, but did the mast of *Adonia* really get under Grunental

High Bridge! What I can tell you is that on her maiden transit of the canal she was delayed by the Canal Authority for three hours on entering the canal at Brunsbüttel, where instructions were given for the top 45cm of the mast to be removed before she was allowed to proceed through the lock. With the anemometer and its supporting pole removed from the top of the mast, she was given permission to proceed with her transit on the understanding that a second anemometer, mounted at the top of the mast, would just fit under the bridges. It really was that tight!

Nearby is the A23 Autobahn Hohenhorn High Bridge, which is followed by the Hochdonn High Bridge that opened in 1919 to carry yet another railway across the canal. Again designed by civil engineer Friedrich Voss, this bridge has in recent years twice been damaged by ship impact – so High Bridge was perhaps not perhaps the most appropriate name for it, being slightly lower than other bridges along the canal. Because of its low height and because of the severe speed restrictions imposed on trains crossing it due to corrosion of the main 122m span, the authorities decided that a replacement was overdue. So in 2006, in another very tricky operation, the old riveted main truss weighing 1,465 tons was disconnected and lowered onto a barge for removal. A new all-welded truss was delivered to the site in the same way, so hopefully no more damage from ship impact. The bridge has very long approach viaducts, and as these were judged to be in reasonable condition they were refurbished as necessary in order to provide decades more service life to the crossing.

The final bridge, just before the locks at Brunsbüttel, is the Brunsbüttel High Bridge that was built in 1983 to carry National Expressway 5 over the canal. Cruise ships using the canal must be of a certain size to pass beneath all eleven of the bridges over the canal. As built, the cruise ship *Norwegian Dream* was too large to transit the canal, as her air draft (the distance from the waterline to the highest part of the ship) was too great to enable it to pass under the bridges. A unique solution was found for this ship – the upper part of the funnel was hinged about its port side and the foremast was similarly redesigned, enabling it to be lowered. In the spring of 1998, six years after the ship was built, she entered the yard of Lloyd Werft at Bremerhaven for the adaptations to the funnel and mast to be made and, whilst there, even more drastic maritime surgery took place. The ship was cut in half and the two halves floated apart. Then a new 40m-long midsection was inserted, the ship having been designed from the outset with the concept of lengthening in mind, making it possible for the company to expand their capacity relatively easily without having to order an entirely new ship. All of the work was completed at the yard within an amazingly short period of three months, the new midship section having been built before the vessel was cut in half. The ship re-emerged from the dockyard with her gross-registered tonnage increased from 40,000 to 50,000 tons and her passenger capacity increased from 1,200 to 2,200 persons. Most importantly, in this final form she could now make passage safely along the Kiel Canal in the knowledge that she would fit under the bridges – but only just! *Norwegian Dream* is probably the largest cruise ship to have regularly transited the canal.

Hochdonn High Bridge (Railway).

Another ship taller than the bridges would otherwise allow was designed from the outset to be able to transit the canal. She is the three-masted, Kiel-based, German Navy sail training barque *Gorch Foch*. Launched from Blohm & Voss yard in Hamburg on 23 August 1958, the telescopic tops to her fore and main masts can be lowered to the level of the mizzen so that she can navigate the Kiel Canal safely, clearing all the bridges.

When the canal was re-engineered in 1895, many stretches of the original Eiderkanal were abandoned. But, just as with abandoned canals in the UK, volunteers are now working to restore and reopen remnant lengths of this old industrial trading route as a leisure waterway. During my time as chairman of the Wilts & Berks Canal Trust, I was fortunate to have made contact with Dr Jürgen Rohweder, who chairs the Eiderkanal Association. Jürgen very kindly provided me with some of the historical data that I have used in this chapter. Having transited the waterway five times, I would describe the experience as being somewhat similar to a journey along an English canal, with cattle grazing along the edge of the water, people waving from their gardens and balconies and cyclists and dog walkers on the towpath. It is just a little surreal experiencing all of this from the deck of an ocean liner.

5

RELIVING THE EXCITEMENT OF TRAVEL ON THE WORLD'S FIRST LINERS

Cunard's *Britannia*. (Author; Cunard)

It is possible in this second decade of the twenty-first century to relive the excitement of travel on the first liners of Cunard and P&O. How, you might ask, can this be possible, as nothing whatsoever remains of those early mid-nineteenth-century ships? Well, I believe that reliving such an experience is conceivable, as well as being educational and great fun, as the following account of two historic British-built ships, one of which was actually built by a civil engineer, will show.

The first Cunard ship was the 1,135-gross-registered-ton paddle steamer *Britannia*, launched on 5 February 1840 from Robert Duncan's yard at Greenock, Scotland, with engines manufactured at Robert Napier's Lancefield Foundry. *Britannia* was the first of sixteen paddle steamers built for the Cunard Steamship Company Limited. She was just 63m long – remarkably, that's less than twice the width of the current Cunard ships *Queen Victoria* and *Queen*

The world's last seagoing paddle steamer *Waverley*. (Robert Coles)

Elizabeth. Britannia bravely transported passengers, mail and freight across the Atlantic for nine years, carrying 640 tons of coal that was consumed at an average rate of 34 tons per day.

Britannia's first westbound departure was from Liverpool on 4 July 1840, taking eleven days and four hours to reach Halifax, Nova Scotia. During the voyage she achieved a top speed of 10 knots. Charles Dickens made one of the ship's first crossings, apparently describing the vessel as being like 'a gigantic hearse with windows'. During the voyage the seas got up to such a height that the passengers feared for their lives. Thankful for their eventual safe deliverance, Dickens, on behalf of all of the passengers on board, presented Captain Hewitt with a pair of inscribed goblets.

The exact end of *Britannia* is not clear, but it is believed that she was sold to the North German Federation Navy in 1849 for conversion to a warship, being renamed *Barbarossa*. Her engines were later removed and she survived as a hulk for many years. Three possible fates have been reported: she was sunk as a target ship by the Prussian Navy; scrapped at Kiel; or broken up at Port Glasgow.

Three gentlemen are credited with starting what today we know as P&O Cruises. In 1822, Brodie McGhie Willcox, a London ship-broker, and Arthur Anderson, a sailor from the Shetland Isles, went into partnership to operate ships between England, Spain and Portugal. Dublin ship owner Captain Richard Bourne joined them in 1835. Together they started a regular steamer service between London and the Iberian Peninsula. Adopting the name Peninsular Steam Navigation Company, they provided services to Vigo, Oporto, Lisbon and Cádiz. In 1837, the company won a contract from the British Admiralty to deliver mail to the Iberian Peninsula and then, in 1840, it acquired a contract to deliver mail to Egypt. A Royal Charter incorporated the Peninsular and Oriental Steam Navigation Company later that year. The P&O Company was born.

The wooden paddle steamer *William Fawcett*, built for a gentleman of that name in 1828 for the London to Cork and Dublin service, was purchased by Captain Richard Bourne in 1832, some three years before he joined forces with Anderson and Willcox. Traditionally regarded as the first P&O ship, this vessel was never officially owned by the company and was broken up in 1845. In the closing months of 1836, the 932-gross-registered-ton wooden paddle steamer *Don Juan*, designed specifically for the Peninsular Steam service, was completed on the Thames at Poplar, London. She operated her first Iberian voyages in 1837, departing the UK on 20 July for her maiden voyage to Gibraltar; her second voyage was to San Sebastian. With the company having just been awarded the Admiralty mail contract, she set out again for Gibraltar on 1 September 1837, this time carrying the mails. The mail was safely delivered to the colony, but less than three hours after leaving Gibraltar on Friday 15 September, on her homeward-bound voyage, disaster struck. In thick fog she took the ground at Tarifa Point at the southern tip of Spain. Fortunately, there was no loss of life and the mails were saved and delivered safely to England. The mail contract remained with the Peninsular Steam Navigation Company but the ship was a complete loss and was abandoned to the elements (the P&O archive has advised that no images exist of *Don Juan*.)

Impossible then to relive the excitement of ocean travel on the *Britannia* or the *Don Juan*, but what do these ships have in common with *Waverley*? All three are paddle steamers, propelled across the ocean by huge rotating side-paddle wheels driven by a mighty steam engine. So it is by voyaging on *Waverley*, the world's last ocean-going paddle ship, that the unique excitement of steam-paddle propulsion can be relived.

The *Waverley* photographed next to *Queen Mary 2* (see page 76) illustrates an interesting size comparison – and remember that *Britannia* was 33ft shorter than *Waverley*.

The *Waverley* story is itself interesting. She was built for the London and North Eastern Railway Company, formerly the North British Railway Company, in 1946 to replace their famous *Waverley* of 1899, which was lost during the Second World War while rescuing troops from the beaches of Dunkirk. A brass plaque on the later ship celebrates the service given by the 1899 *Waverley* in both world wars – as a minesweeper in the First World War and as a transport ship in the Second World War. Both the 1899 and the 1947 *Waverley* were built in the Glasgow yard of A. & J. Inglis Limited, later to become part of the Harland & Wolff shipyards on the Clyde. Sadly, shipbuilding on the Clyde, as with that at many other locations across Great Britain and Northern Ireland, is now but a shadow of what it was when the UK was one of the greatest shipbuilding nations in the world.

On a misty day in October 1946, Lady Matthews, the wife of London and North Eastern Railway (LNER) chairman, launched the last paddle steamer to be built for Clyde services. The new steamer was towed to Victoria Harbour, Greenock, where Rankin & Blackmore Limited fitted the double-ended steam boiler and the triple-expansion diagonal steam engines. She was delivered into LNER service for the 1947 season sporting the company's standard red funnel with white band and black top, but was only able to retain this distinctive colour for one year, as at the end of 1947 all of the privately owned railway companies in Britain were grouped into the massive nationalised British Transport Commission (the commission adopted a yellow and black funnel colour throughout the newly created national fleet). *Waverley*'s primary role was to extend the LNER passenger service from Craigendoran Pier cruising Loch Long and Loch Goil to the villages of Lochgoilhead and Arrochar. In addition she undertook her share of the many excursions from Craigendoran to Clyde Coast resorts of Wemyss Bay, Rothsay, Largs, Troon, Ayr and Brodick on the Isle of Arran.

In November 1951 the British Transport Commission set up a Scottish shipping division, the Caledonian Steam Packet Company Limited, transferring into it *Waverley* and the other two Craigendoran ships *Jeanie Deans* and *Talisman*. In 1973, this

Size comparison between *Queen Mary 2* and *Britannia*.

fleet and much of the fleet of David MacBrayne Limited were amalgamated to form Caledonian MacBrayne Limited. Sadly, the new company, which today still provides most of the Scottish Islands' services, immediately announced its intention to dispose of *Waverley*. However, instead of scrapping this now unique and historic vessel, Caledonian MacBrayne offered her to the then newly created Paddle Steamer Preservation Society for the princely sum of £1. Originally, it was thought of as being only suitable for static preservation but the more ambitious members of the society (and I count myself amongst that number) turned our minds to actually operating the vessel. Recognising that to do so would require huge sums of money, appeals for funding were made to local authorities, tourism bodies and countless commercial companies. Somehow we raised the money, enabling the steamer to operate her first cruise under the Waverley Steam Navigation banner in May 1975. Now, with National Lottery support, each year *Waverley* undertakes an ambitious programme of cruises around UK coasts, providing many with the opportunity of experiencing the beating heart of the ship – the mighty steam engine that turns the giant paddle wheels located on either side of the vessel, exactly like those engines and wheels that once drove the Cunard *Britannia* and the P&O *Don Juan*. The

The launch of *Waverley*, 2 October 1946. (Paddle Steamer Preservation Society)

Triple-expansion diagonal steam engine of *Waverley* built by Rankin & Blackmore.

majesty of engines of this type is not easy to demonstrate in a still photograph but with the same sounds, the smell of steam and oil and the same paddle-driven motion, it really is very easy to imagine oneself aboard one of those first early steamships powering away from the quayside on a voyage of excitement and discovery.

Well, that takes care of the propulsion experience, but what of revisiting the type of accommodation provided on those early Cunard and P&O steamships. For that, we have to turn to a ship that was built at exactly the same time as the Britannia class of ships and which happily survives today, sitting in the dry dock in which she was built in 1843. I am referring to Brunel's SS *Great Britain*, which can be visited in Bristol City Docks in South West England. She is now fully restored as a multi-award-winning museum ship where you can experience first hand the sort of passenger accommodation and entertainment spaces available in those early steamships. The story of this Bristol-built ship has been dramatic and is well worth telling. The SS *Great Britain* was designed by one of the greatest civil engineers of all time – Isambard Kingdom Brunel. Brunel had a vision to extend his London to Bristol Great Western Railway westward from Bristol to New York. In 1837, three years before the launch of Cunard's *Britannia*, he launched the wooden paddle

steamer *Great Western* from William Patterson's yard in Bristol for the Great Western Steam Ship Company. She completed her first crossing to New York in April 1838. Up until that time the US dominated the North Atlantic passenger trade, running fast sailing packets of 300 to 500 tons providing a reliable service with departures on pre-published dates but, of course, arrival times very much depended on encountering suitable winds on the voyage.

With *Great Western* of 1,320 gross-registered tons, Brunel proved that it was possible for a steamship with paddle wheels to make regular voyages back and forth across the Atlantic without running out of coal, a significant worry among the travelling public at the time. Her speed, reliability and comfort quickly eclipsed the sailing packets. As confidence grew in *Great Western* so did Brunel's ambition. His next step would be to build a bigger, better and faster sister ship, but she would still be constructed of wood and driven by paddles and he intended to name her the *City of New York*. But that was not quite how his ideas developed. In 1838, before work had actually started on the new ship, a little paddle wheeler called *Rainbow* visited Bristol, an event that fundamentally changed Brunel's design concept. So intrigued was Brunel with the fact that she was built of iron that he sent his colleagues Captain Christopher Claxton and

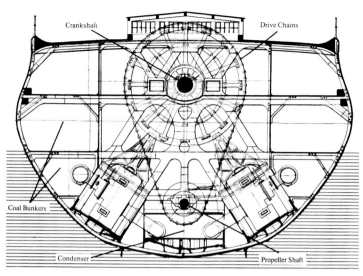

Above: Paddle steamer *Great Western* of the Great Western Steamship Company. (SS Great Britain Trust)
Right: *Great Britain* engine detail (drawing by Weale). (SS Great Britain Trust)

William Patterson to travel on it to Antwerp and, based on their favourable report, he decided that his great new ship must also be built of iron. However, finding that no company was prepared to tender for such an unusual vessel that would include larger paddle wheel engines than had ever been built before, Brunel decided that the Great Western Steamship Company would itself build the ship. He first built an engine factory alongside the harbour in Bristol and then a huge dry dock in which to construct the ship. The keel was laid down as soon as the dock had been completed and wrought-iron plates arrived in barges from Coalbrookdale in Shropshire; soon the Great Western Dockyard became the first place in the world where all of the processes associated with modern iron ship-building came together on one site.

Ten months into the build a second change of plan was to take place, perhaps even more radical than the change from wood to iron. This change came about following a visit to Bristol of another small experimental vessel. In May 1840, the *Archimedes* gave demonstrations of her novel screw propeller designed by Francis Pettit Smith. Brunel was so fascinated by this new means of propulsion that he chartered the vessel for six months so that he might carry out experiments of his own. As a direct result of those experiments he became convinced that he must alter the motive power

of the vessel he was now building. He summarised for the company board the arguments for change of propulsion for the new vessel, stating that as screw-propulsion machinery was lighter in weight than paddle-wheel propulsion machinery it thus improved fuel economy, took up less space – making more room for cargo – and, being placed lower in the hull, improved the ship's centre of gravity – making it more stable in heavy seas. He opined that the depth of immersion of a paddle wheel is constantly changing, depending on the ship's cargo and the movement of waves, while a propeller stays fully submerged and at full efficiency at all times. Brunel's arguments proved persuasive and in December 1840 the company agreed to adopt the new technology. The decision, however, became a costly one and set the ship's completion date back by nine months. On the drawing board, his engine had already been designed to turn a paddle shaft and side-wheel paddles, so these were removed and the engine turned through 90 degrees, allowing the crankshaft to directly drive an 5m-diameter gear wheel and, via a drive chain, a 2m-diameter gear wheel mounted on the propeller shaft. The gear wheels and drive chain were 96cm wide. A full-size replica of Brunel's original engine has been installed in the restored ship. This engine, although not powered by steam, rotates as if it was and is complete with realistic sound effects. *Great Britain* was

Great Britain rudder and propeller detail. (Derbyshire Dave via Wikimedia Commons)

Great Britain apparently afloat in her building dock. (SS Great Britain Trust)

Launch of *Great Britain*, 19 July 1843. (SS Great Britain Trust)

the first ship of any note to be fitted with a balanced rudder where its line of balance ran down through the centre; this meant that it, and the ship, could be turned in the water far more easily than with a conventional type of rudder that rotated about its trailing edge. The propeller, designed by Brunel, was the first large one ever built.

Access down into the dry dock by stairs or by lift is part of the museum experience, enabling the visitor to closely examine the detail of the rudder, propeller and the underwater shape of Brunel's hull. The hull was constructed of 2m x 1m wrought-iron plates riveted to metal frames. The plates overlap each other in 'clinker' style, which is estimated to give 15 per cent more strength to the hull than if the plates had been laid edge to edge. The plates are closely riveted and the hull has always been remarkably watertight. The size and construction details of the Great Western Dry Dock itself, which was specifically built in order to construct the ship within, can also be appreciated. It is now completely weatherproof, as it has been provided with a glass roof built on the waterline of the ship. A modern air-drying system designed to remove 80 per cent of humidity to help stem further corrosion of the ship's hull has been installed. The glass supports some 15cm depth of water up through which the upper part of the hull can be seen and, cleverly when viewed from above, *Great Britain* is once more 'afloat' in her building dock.

She was one of the first ships to be provided with bulkheads to divide the interior space into watertight compartments sealed by watertight doors. Five bulkheads ran across the inside of the hull and two longitudinal ones ran either side of the engine room, adding great strength to the hull. The masts were rigged with wire rope instead of traditional hemp, which meant that the rigging would remain taught whatever the weather. Only one mast went down through the ship to the keel; unusually, the other five masts were hinged at the weather deck. The hinges allowed a certain amount of movement at the base of each mast in order to help the fitting of the

wire rigging. She was the first large ship to be built with a clipper bow, designed for speed and able to cut through the water, this in contrast to the traditional rounder, bluffer style of bow that punched its way through the waves. The bow and underside of the ship have very fine lines, whereas above the waterline, she bulges perceptibly in order to provide as much accommodation as possible.

Brunel's wrought-iron ship, then the biggest in the world, was launched on 19 July 1843 by Prince Albert, husband and consort of Queen Victoria, before thousands of onlookers who had gathered to witness the spectacle. Those viewing from inside the steamship yard each paid 5*s* for the privilege. In celebration, church bells rang out all over Bristol and a huge fireworks display at Bristol Zoological Gardens took place that evening, which included a depiction of the ship set in a sea of blazing sparkler fountains. A period of fitting out followed the launch before the vessel was ready to take to the seas. Unfortunately, the ship was too large to pass out through the locks joining the Floating Harbour to the River Avon. Although this had been known about for some time, apparently the harbour authority had been tardy in making the necessary modifications to the locks and so the ship's departure was delayed until November 1844.

Setting off on her maiden voyage, *Great Britain* was assisted down the River Avon by steam tugs as she passed the site of Brunel's

Temporary protective mattress built around the ship in Dundrum Bay to help prevent damage from winter storms. (SS Great Britain Trust)

yet to be completed Clifton Suspension Bridge. Another 126 years were to elapse before *Great Britain* was to pass beneath the completed bridge. Making her way down the Bristol Channel, she ran into heavy seas near Lundy Island but weathered the storm just as she was designed to do. Brunel designed *Great Britain* as a 'sail-assisted steamship'. In addition to steam engines and a propeller she carried six masts with a special 'schooner rig', which was necessary because when the engine drove the ship forward the virtual wind direction was from ahead and this did not suit conventional 'square rig' sail patterns. In storm conditions, when it was prudent not to raise sails, the ship proceeded ahead on engines only. Because of problems associated with the transit from Bristol to the open sea for such a large ship, which included navigating around the notorious River Avon Horseshoe Bend, the Great Western Steamship Company somewhat reluctantly decided to move its operation to Liverpool.

Disaster struck the ship on her fifth voyage when travelling westbound from Liverpool to New York on 22 September 1846. She had embarked 180 passengers and a substantial amount of cargo, leaving Liverpool at 11 a.m. Later that dark and wet night, having spotted ahead off his starboard quarter what he took to be the Chicken Rock Light on the Calf of Man at the extreme southern end of the Isle of Man, Captain Hosken knew that he needed to continue westward for some distance past the light before making the turn north that would have taken the ship between the Isle of Man and the Irish coast. At least, Captain Hosken *thought* he had spotted the Chicken Rock Light to the south of the Calf of Man. Unfortunately, having completely missed the Isle of Man lighthouse, the light ahead on his starboard quarter was in fact St Johns Light on the Irish coast. Believing it to mark the southern tip of the Isle of Man, he set a course to pass it on his starboard side and then continue further west before turning north. As he was soon to find out, there was no water to the west of St Johns Light, only solid land, and so, in poor

visibility and at some speed, he ran hard ashore on the beach of Dundrum Bay in County Down, just to the south of St Johns Light. Captain Hosken had in fact gravely underestimated the speed at which he had been running. First the compass was blamed and then the charts, but the fact remained that *Great Britain* was a very long way from where Hosken thought she was. Thankfully, no one was seriously injured, but all on board spent the rest of an agonising night with waves breaking against the sides of the ship. At daybreak, with the passengers safely disembarked, inspections revealed that, although leaking slightly, the strongly built vessel had sustained very little structural damage, but she was firmly ashore and lying between two outcrops of rock.

Due to the coming winter there was no question of her being refloated straightaway. When news reached Bristol, Brunel expressed dismay and fury at what had happened to his beloved ship and made arrangements to journey to Ireland, arriving at Dundrum Bay early in December. Measures were immediately taken to protect the vessel from winter storms using a combination of faggots, rocks and timber. A contemporary watercolour by an unknown artist shows, nearing completion, the protective mattress built around the ship. The following spring, the ship was ingeniously lifted just

Poster advertising passage to Australia aboard *Great Britain*.
(SS Great Britain Trust)

clear enough of the sand for boilermakers to get beneath the hull to repair her damaged plates. Raking timber posts were driven into the beach on either side. Ropes were run over the tops of these posts and under the hull, and to the ends of the ropes 50-ton boxes of sand were attached. As the boxes descended slowly down the posts the ship lifted, allowing supporting timbers to be pushed into the cavity below the hull creating the space for the repair work to be

undertaken. A trench was excavated between ship and sea, and at each high tide she was inched back to deep water by a combination of ground-pulling winches and tow lines from HMS *Birkenhead* and HMS *Scourge*. On 27 August 1847, she was once again floating free and was towed back to Liverpool for the completion of permanent repairs in the shipyard of Fawcett, Preston and Company. Although the ship had been saved and had a long working life ahead of her, the Dundrum Bay grounding marked the end of the Great Western Steamship Company. *Great Britain* was under-insured and the cost of the salvage operation and repairs could not be met, so both *Great Britain* and the older *Great Western* were put up for sale.

In December 1850, *Great Britain* was bought by Gibbs, Bright & Co. of Liverpool, a company that had an interest in running services to Australia, where recent discoveries of gold had stimulated intense interest. A new engine was fitted to the ship, with steam being supplied at double the original pressure from new boilers connected to twin side-by-side funnels. The number of masts was reduced from six to four and, when under sail alone, a new three-bladed propeller could be de-clutched from the propeller shaft to reduce underwater drag. The accommodation was also radically altered, with capacity being increased to 730 passengers of whom fifty were first class. This increase was achieved by adding additional bunks to some cabins and a new deckhouse was constructed on the weather deck, which ran almost the entire length of the ship. By 1852 all was ready for the Australia service, but a first trial voyage was made to New York under command of Captain Robert Mathews. That voyage was a success, with the westbound passage being made in thirteen days at an average speed of just over 9.7 knots. Soon after her return to Liverpool, she set out on 21 August 1852 with 630 passengers on board on her first voyage to Australia, reaching Melbourne eighty-three days later. Her homeward passage was via the Cape of Good Hope, arriving back in Liverpool in April 1853. She carried gold as part of her cargo, having been equipped with

Longitudinal section through ship showing propeller lifting frame. (SS Great Britain Trust)

six eight-pounder guns for protection. A break in the Australia voyages occurred in September 1855 when, just as Cunard's *QE2* and P&O's *Canberra* were many years later requisitioned to take troops to the Falklands War, so *Great Britain* was requisitioned by the British government to transport troops to the war in Crimea. She was again called up for military duties in 1857 to transport troops to Bombay during the Indian Mutiny. Her new owners were, however, not satisfied with her performance on the Australia service and,

through further refits, her two funnels were replaced with a single small one; three square-rigged masts replaced the previous four; a large bowsprit was fitted and another unique feature was added to the ship. A mechanism was installed to uncouple the propeller from the shaft and then the whole propeller assembly was winched up into a cavity within the ship. This lifting frame was cutting-edge technology, allowing the ship to sail when the wind was favourable without the propeller dragging in the water and slowing her down.

Great Britain abandoned in Sparrow Cove, Falkland Islands, 1968.
(SS Great Britain Trust)

She was effectively now an 'auxiliary steamer' with engine and propeller as a back up to sails. In this guise she settled down to a further twenty years of steady passages between Liverpool and Melbourne carrying the forebears of probably hundreds of thousands of present-day Australians.

By 1876, her owners announced that her passenger-carrying days were over and she was laid up at Birkenhead and offered for sale. In 1882 she was acquired by Antony Gibb, Sons & Co., which removed all of the passenger accommodation, engines, boiler and funnel. Three new cargo hatches were provided through the upper deck and for added strength a belting of thick, caulked pitch-pine cladding was added to the outer hull between high and low loading marks. So began a new career as a windjammer carrying South Wales coal from Penarth to San Francisco under the command of Captain Henry Stap. These were long, slow voyages, each round-trip lasting over twelve months. She completed two voyages, numbers 45 and 46, successfully, but it was on her forty-seventh voyage, when outward bound from Penarth, that she finally came to grief while attempting to round Cape Horn. Rounding the cape in a sailing ship from east to west was always difficult because of the strength of the prevailing westerly winds. When almost off the Cape, she began to get into serious difficulties, with seas breaking right over her. The decks were leaking and the old hull, deep-laden with coal, was felt to be straining in the mountainous seas. For almost a month they struggled to round the Cape, but, after the cargo shifted and the fore and main topgallant masts were both carried away, Captain Stap decided to turn and run before the gale for shelter in Port Stanley in the Falkland Islands, where she arrived on 26 May 1886. *Great Britain*, worn out and damaged, was never to sail again.

After over forty years at sea, in 1887 she was declared a 'hulk' and subsequently used by the Falkland Islands Company as a floating wool and coal storehouse in Stanley Harbour. In 1937, almost 100 years after Brunel sat down to design her, with rotten decks and leaking hull, she was declared uneconomic to operate even as a store ship and was towed 5.6km out of the harbour and beached at lonely Sparrow Cove. Explosives were used to make holes in the iron hull so that the ship would remain on the seabed whatever the state of the tide. But that was not the end of *Great Britain's* usefulness. HMS *Exeter*, having suffered heavy damage and loss of life during the Battle of the River Plate on 13 December 1939, limped into Port Stanley where her crew, desperate to make their ship seaworthy, wrenched and cut iron plates from the hull of *Great Britain* and riveted them to their own damaged hull, enabling the ship to safely return to England and to eventually serve in battle again.

No serious attention was paid to *Great Britain's* historical importance in maritime history until 1967 when a British naval architect, Dr Ewan Corlett, wrote to *The Times* newspaper. Following Dr Corlett's intervention, a major salvage operation was got under way in 1970. After some temporary patching of the hull, which permitted her to float, the semi-submersible ocean-going

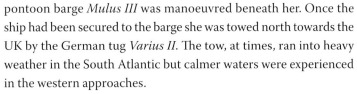

Great Britain in the western approaches aboard pontoon barge *Mulus III*. (SS Great Britain Trust)

Pleasure steamer *Balmoral*. (Robert Coles)

pontoon barge *Mulus III* was manoeuvred beneath her. Once the ship had been secured to the barge she was towed north towards the UK by the German tug *Varius II*. The tow, at times, ran into heavy weather in the South Atlantic but calmer waters were experienced in the western approaches.

Back in 1970, a pleasure-steamer service was still operating from Bristol and other ports in the Bristol Channel. On consulting the sailing timetable for the day that *Great Britain* was due home, I noted that a steamer was scheduled to depart from Hotwells Landing Stage, Bristol, at 8.45 a.m. for a cruise down the Bristol Channel calling at Clevedon, Weston, Ilfracombe and Lundy Island. Having checked the expected arrival time of *Great Britain* in the upper Bristol Channel, I booked a ticket for Weston-super-Mare aboard the White Funnel pleasure steamer *Balmoral*. Towards the end of my excursion, *Balmoral* approached within hailing distance of the tow, after which I disembarked the steamer at Weston, highly

Post Office special cover celebrating the end of Voyage No. 47.

Great Britain's return to Bristol, 5 July 1970. (SS Great Britain Trust)

Post Office special cover celebrating the re-docking of *Great Britain* on 19 July 1970.

delighted to have actually witnessed the end of *Great Britain*'s Voyage No. 47, which had, unbelievably, begun in February 1886 – eighty-four years earlier!

A couple of hours after my encounter, *Great Britain* arrived at Avonmouth Docks to be greeted by a salute from the siren of every ship in the port and to enormous interest throughout the nation. To celebrate the historic end to Voyage No. 47, a commemorative First Day Cover was produced. Reliving her days as a Royal Mail ship, *Great Britain* actually carried this special mail from the Falkland Islands to Bristol. At Avonmouth Docks the ship, still on the pontoon barge, entered dry dock where a certain amount of further temporary patching of the hull was undertaken. Then with all straps and braces between ship and barge removed, with the sea cocks open on the barge allowing it to remain on the floor of the dock as water rose within, *Great Britain* floated free.

A few days later, now under the control of two Bristol tugs, she was once more under way in her home waters, making stately progress up the River Avon to her birthplace in the centre of Bristol.

An estimated 100,000 people lined the banks of the river to witness Brunel's bridge and his ship meet for the first time. The bridge, of course, had not been built at the time of *Great Britain*'s departure from Bristol (I was somewhere in that crowd with camera held aloft and a sizeable lump in my throat). The ship was moved through the river lock across Cumberland Basin and into the Floating Harbour, where she was turned and moved astern to tie up at Y Wharf right opposite the Great Western Dry Dock where she was built. She waited at this wharf for a high enough spring tide to get her through the shallow and narrow entrance into the dry dock. By pure and extraordinary luck, the first date that provided the necessary tide for this final move of the ship was 19 July, the precise anniversary of her launch from that very dock in 1843. A lucky date, too, as it was a rare day in the Duke of Edinburgh's diary where his presence could be arranged at short notice. Thus, on 19 July 1970, the husband of Queen Elizabeth II was on board for the re-docking, just as 127 years before on 19 July 1843, the Prince Consort, husband of Queen Victoria, had launched *Great Britain* (so a right royal

Inside the dry dock showing its glass roof, which supports 150mm of water. (SS Great Britain Trust)

send-off and return for this very special ship). Luck was also with me on 19 July. I just happened to be passing through Bristol on that summer evening and stopped to investigate why crowds were gathering. Happily I witnessed the whole unpublicised event as *Great Britain* was carefully manoeuvred back into the Great Western Dry Dock. And, to celebrate the re-docking, another First Day Cover was issued. (Note the image on the postage stamp on page 87; it is of the very last transatlantic passenger liner to be built in Britain

– *Queen Elizabeth* 2. She was completed at John Brown's famous shipyard in Glasgow in 1967.)

Since that July day in 1970, I have, as a member of the Great Britain Trust, regularly visited *Great Britain* to follow the painstaking restoration activities. So what will you be able to see if you visit the ship in Bristol? You will be able to wander the decks, public rooms, entertainment spaces, cabins and working areas of the ship and relive, as I suggested at the outset of this chapter, the sights, the

First-class promenade saloon.

sounds and the smells of life on board a ship of the age of those first Cunard and P&O ships *Britannia* and *Don Juan*. Your experience will begin with the purchase of a ticket for a passage to the Port of Melbourne in Australia, a ticket, incidentally, that can be used for as many 'voyages' as you wish in a period of twelve months if you sign up for Gift Aid at the outset.

You will first be invited to visit the Great Western Dry Dock, which is accessed from the quayside either by stairs or by lift. Down inside the dock, dry whatever the weather outside, you can examine the results of Brunel's civil engineering expertise in the construction of the dock itself and marvel at his shipbuilding skills by being able to examine every underwater detail of *Great Britain*'s hull. After the dry dock, you can visit the onshore museum at quayside level where you can spend as much time as you like. Exhibits, video presentations and hands-on interpretation cover the complete life of the ship. From the upper floor of the museum, access is gained to the gangway that leads directly onto the ship, boarding the vessel for your 'voyage' to Melbourne via the main weather deck. Of note on this deck are the livestock pens – something we don't think about today, taking for granted refrigerated meat storage facilities on modern ships. From deck houses on the weather deck, compan-

ion-way stairs lead down to the promenade deck, at the forward end of which is located steerage accommodation. The upper part of the engine room and galley are located midships and further aft is located the first-class promenade saloon with first-class cabins on either side.

The promenade saloon is a completely enclosed wide-open space used by first-class passengers to 'promenade' – to meet and socialise with fellow passengers without getting wet or windswept on the upper deck. It is lit with natural daylight by skylights set into the weather deck above. Light wells beneath each skylight transmit daylight to the deck below. At the after end of this saloon, there is a window seat across the full width of the ship where passengers would sit and watch the ship's wake streaming out behind and, on either side forward, cosy private rooms were set aside for female passengers.

Great Britain was designed as a deluxe passenger ship for the Bristol to New York service, so the first-class cabins opening off of the promenade saloon provided a high standard of privacy and comfort, if a little more basic than those we enjoy today. Note the smoky candle lamp (see page 90); fear of fire at sea meant that the ship's officers controlled the use of candles very strictly and all lights had to be out by 10 p.m. Several of the cabins and other compartments are set out with 'speaking' life-size tableaux depicting events from the actual logs of voyages. For example, travelling in berths 99–100 in 1875, Georgina Bright cares for her sick children while her nephew lies dying of typhoid in the opposite cabin. Ships Surgeon Samuel Archer tries, in 1860, to save a seaman's injured hand and, in berths 97–98, Mr Jones has set up a barbers shop where he is debating with his customer Mr Dearlove the issue of slavery in America's southern states – a hot topic in 1862.

Moving forward along the promenade deck, it is the upper engine room that next comes into view. The sound effects and smells here are amazing; stokers viewed through the metal bars of the engine-room floor can be heard shovelling coal three decks below as they

First-class cabin.

Part of the full-size reconstructed engine.

First-class saloon set up for dining. (SS Great Britain Trust)

Steerage-class dining arrangements. (Keith Bennett)

First-class saloon set up as an entertainment space.

Waverley at Ilfracombe, North Devon.

toil to feed the hungry steam boilers. Steerage accommodation at the forward end of the promenade deck was described by an immigrant passenger as being rather cramped, but was not so bad as long as you got on with your fellow passengers. Allan Gilmore described his State Room, as the company called it, as a very confined and dark space but fairly well ventilated. It contained four berths and the space between them for the purpose of dressing was just ½m wide by 2m long, allowing only one person to dress at a time. Even in this small area, some items of personal luggage had to be stored. The ship's main galley, situated between the engine room and steerage, often had to cater for 600 persons. On a typical voyage, it was staffed by ten cooks, two bakers, two butchers and a storekeeper. On passage to Australia, the ship could be continuously at sea for up to sixty days, so all food and freshwater needed on the voyage had to be loaded on board before the ship left Liverpool. There were storerooms on board for dry foods, vegetables, cheese and cooked meats. An ice room kept meat fresh for the first week, after which salt meat was supplemented with fresh meat from the livestock provisions on the weather deck.

First-class passengers ate very well and were served plenty of fresh meat and soft bread in the splendid dining room located on the saloon deck below. Steerage-class passengers had to take it in turns to fetch food from the pantry for a group of passengers, take it to the galley to be cooked and, when ready, collect it from the galley and serve it to their group on wooden benches ranged around the outside of their sleeping accommodation.

Between meals, the first-class saloon also served as the main entertainment space for first-class passengers, who largely made their own entertainment on board. Concerts, amateur theatrical performances, charades and even mock trials were all very popular. In addition, conjuring displays, bible-reading groups, language lessons, volunteer military drills and even pillow fights all helped avoid passengers getting bored at sea – a little different to the entertainment provided at sea today.

So there we have it. It really is possible, in this second decade of the twenty-first century, to relive the excitement of travel on those first Cunard and P&O steamships, to smell the steam and oil, to experience the unique nature of paddle-wheel propulsion and to wander around the decks and take in the very different levels of appointment provided to first- and steerage-class passengers of that bygone age.

www.waverleyexcursions.co.uk
www.ssgreatbritain.org

Great Britain dressed overall. (SS Great Britain Trust)

6

RESTORING NELSON'S WATERWAY

BREATHING LIFE INTO A NORFOLK BROAD

Civil engineering is not just about the building of mega-structures. This story is about the restoration of the body of water upon which it is reputed that, as a boy, Admiral Lord Nelson, learned to sail.

During my time as Civil Engineer to the [Norfolk] Broads Authority I led a team that returned a large 'broad' back to health. Perhaps I should at the outset, point out to American readers that the word 'broad', from our shared language, means something entirely different on either side of the pond. In the East of England, a broad is a large body of water or lake. Following its restoration, the broad was once again able to contribute to the growth of tourism in the area but, at the same time, it required careful management in order to protect what is recognised internationally as a valued and fragile wetland habitat. The story will, I hope, illustrate for you that civil engineering is not just about steel and concrete.

The light-shaded area of the map (see page 94), straddling the counties of Norfolk and Suffolk on the east coast of the United Kingdom, shows the extent of the Broads Authority Executive Area, headed from its inception in 1979 until 2001 by Professor Aitken Clark, who sadly passed away in 2010. During those early years, steered by Aitken, the area was an associate member of the national parks family but today it is, in its own right, one of Britain's fully operational National Parks.

The body of water associated with Nelson is Barton Broad, the second-largest broad in East Anglia. It is located on the River Ant at the northern end of a web of navigations linking the hinterlands of Norfolk and Suffolk, including the city of Norwich, to the North Sea through the ports of Great Yarmouth and Lowestoft. The system is tidal but the variation in water level at Barton Broad is only about 15cm. At the coast, the water is salt but soon reverts to a brackish nature the further one travels up the Rivers Bure, Yare and Waveney, and before long its nature becomes fresh.

Horatio Nelson's connections with Norfolk are well documented. He was born the son of a country parson in 1758 at Burnham Thorpe

Aitken Clark OBE
1936 - 2010

A tribute to the late Aitken Clark OBE.

A busy scene at Norwich City Quay in the 1950s. (Broads Authority)

in North Norfolk and joined the navy as a captain's servant in 1770 at the tender age of 12. He was killed in 1805 on board HMS *Victory* during the crushing defeat of the French and Spanish fleets at the Battle of Trafalgar. The most well-known memorial to Nelson is, of course, Nelson's Column in Trafalgar Square, London, but in the grounds of Norwich Cathedral another memorial exists to this son of Norfolk.

So, just how was Nelson's Broad formed? It is, perhaps, hard to imagine that the entire area of the Norfolk and Suffolk Broads was dug by hand for the removal of peat as a fuel and that, surprisingly, it was not until the 1960s that the man-made origins of the Broads was first understood. The adjoining rivers were used to convey the peat to nearby cities and towns, and over time the rivers became joined to the flooded peat workings, which we now know as Broads. Most of the peat digging in Norfolk in the fourteenth century was organised and recorded by the monks at St Benet's Abbey, the ruins of which, including that of a much later windmill, still stand alongside the River Bure navigation.

For hundreds of years, freight was carried throughout this navigation system by the traditional, single-sailed Norfolk wherry. Prior

to the First World War, scores of these vessels sailed the waters of the Broads. Since much of the area in those days was unreclaimed marshland, with few established roads, transport by water assumed a much greater importance than it did elsewhere in the country. The Broads waterways remained important for the carriage of freight right up until the late 1950s, as the busy scene at Norwich City Quay confirms (above). The very last cargo vessel to unload at Norwich did so in 1989. As has happened to so many inland ports throughout the UK, this entire area has, over recent years, been redeveloped for leisure and tourism.

As the twentieth century progressed, the Broads waterways became more and more popular as boating holiday destinations. Many visitors travelled to their holiday by the once extensive network of railway lines serving that part of Britain. Broads holiday advertisements were commonly seen at railway stations and in the compartments of train carriages, offering holidays at just £4 per person per week. Newspapers and magazines regularly carried advertisements from the two famous boat-hire brands of Blake's and Hoseasons (I particularly remember seeing such advertisements in the *Radio Times*).

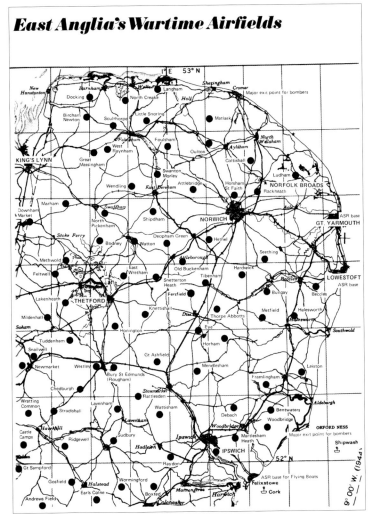

East Anglia's Wartime Airfields

Map indicating the Second World War airfields of East Anglia. (Taken from *The 1,000 Day Battle* by James Hoseason)

his father had fallen ill. He was never to return to engineering, instead spending more than half a century building up his father's Norfolk and Suffolk holiday business into a multi-million-pound company. Jimmy's other passion was flying. He co-founded the Waveney Flying Group in 1960 at Seething Airfield in Norfolk and was involved in establishing the 2nd Air Division Memorial Library at Norwich to commemorate the American airmen who were stationed in the region during the Second World War. He wrote an authoritative book on the subject entitled *The 1,000 Day Battle*. Because of its close proximity to Europe, dozens of airfields were built in East Anglia and every dot on a map recorded in his book marks the position of an operational wartime airfield, many being operated by the US 8th Army Air Force; it's a wonder there was any room left at all for the Broads waterways.

Late nineteenth- and early twentieth-century photographs suggest that Broadland shallow lakes were then in excellent condition, displaying extensive beds of lilies and fringing reed swamp. But this idyllic scene was about to change and not, I am sorry to say, for the better. Lakes such as Barton Broad began to silt up and the once gin-clear water was to become cloudy and lifeless. Instead of a coherent fringe of reed swamp, remnant clumps were all that remained and at the edges of the lake the water became noticeably shallow. The banks beyond were often backed by ploughed fields, intensively managed for maximum food production using the most modern agricultural practices of the day. The main body of the water was no longer graced by lily flowers or submerged plants, down whose stems damselflies once climbed to lay their eggs. All of this had gone. The water was no longer clear but was laden with mud or algae, and at the surface the paint-like scum of stranded algal blooms harboured toxins capable of killing farm animals or domestic pets unlucky enough to ingest these organisms. Out in the middle of Barton Broad, a once beautiful island had quite literally disappeared. Protecting reed had died back exposing the island's

When I joined the Broads Authority Team, James Hoseason, the son of the founder of Hoseasons, was a senior member of the Broads Navigation Committee. James and I became good friends, sharing common roots in the profession of civil and structural engineering. One day whilst at work in his London engineering office, Jimmy was called home to help out with the family business because

Thurne Mill, Norfolk. (Broads Authority)

banks to wind-generated waves and those created by passing boats. Not a pretty situation! Although very much reduced in value as an aquatic habitat, the lake struggled on as a leisure resource. It was home to the Nancy Oldfield Trust, which pioneered the provision of specialised facilities to enable people of limited mobility to get out onto the water under sail. As a regular user of Barton Broad, the Trust was in a unique position to be able to record the steady decline of the environment, in both the depth of water for navigation and the quality of the water for wildlife.

With many stunning views across Broadland, like that at Thurne Mill, the seriousness of the problem was not often recognised by short-stay visitors – perhaps just as well for the sake of the local businesses trying to make a living in a very difficult environment. The condition of Barton Broad, like so many shallow lakes around the world, had been greatly changed by the activities of increased human populations and the demands that those populations made on the environment. With larger concentrations of people, the use of cesspits at each individual dwelling was no longer an acceptable means of dealing with human waste. This led directly to the development of municipal sewage treatment works with their concentrated outflows into local rivers. Also, greater populations demanded more food. Chemicals, the long-term effects of which were not fully understood, were poured onto the land in vast quantities to boost crop production. By the natural process of rainfall, these chemicals soaked into the ground and were washed into streams feeding into the lake. Decades of nutrient enrichment from these two principal sources triggered a chain of ecological events that led to clouding of the water with algae, complete loss of water plants and a severe decline in wildlife and fishery. Barton Broad became murky and lifeless, with huge deposits of black, oozy mud making the water extremely shallow and difficult to navigate. Core samples taken down through these sediments revealed the seriousness of the situation. Typical core results indicated that prior

to 1400, basin peat was present. From 1400, the time when peat was first dug for fuel, up to about 1924, a peaty marl was found in the core sample. After 1924, the year that the first sewage works began to discharge into the Broad, the sample tube was full of a thick black silt, and beyond 1975 an even thicker oozy sludge. Laboratory analysis indicated a corresponding increase in the amount of phosphorous concentrations within the sediments, particularly from about the middle of the century onwards, this being a period of a large local population growth. It was soon realised that key to the degradation of the water body were the flows of nitrogen and phosphorus into the water – phosphorus from sewage outfalls and nitrogen from agriculture processes.

The water column was filled with phytoplankton (microscopic algae) only 10 microns across but occurring in densities of up to 1 million per cubic centimetre, therefore adding to the blackness of the water. Submerged plants starved of daylight died out and a blue/green algal-dominated, extremely eutrophicated Broad was the result (eutrophication is essentially a fertilisation of the water through nutrient enrichment). Then, as the algae died in its billions, it settled to the lake bottom, building up levels and reducing the depth of water for navigation.

A survey was undertaken recording navigable depths immediately before the restoration of Barton Broad. The useable area of water, particularly for sailing vessels with keels (indicated by the dark blue and black areas on the diagram on page 98) was basically reduced to a narrow central strip and even this area was dotted with lots of shallow spots. Based on fifteen years of previous experience with smaller broads, the Authority decided to embark on a programme of restoration of Barton Broad, a restoration that would be undertaken in four parts. Firstly, the continuing discharge of phosphates from sewage treatment works would need to be stopped or drastically reduced. Secondly, the nitrate runoff from farming operations would require a similar substantial reduction. Thirdly, because of the huge store of phosphorus and nitrate elements already trapped within the oozy sediments of the Broad, these sediments would need to be removed from the water. Finally, significant parts of the water body were to be bio-manipulated – more about that process a little later!

Phosphorus was the key element to tackle, since by far the greatest amount came from the concentrated outfalls of two particular sewage treatment works; smaller quantities came from the general catchment area. The local water and sewage company Anglian Water played its part by installing phosphate-stripping equipment at its sewage treatment works directly affecting the Broad. They also diverted effluent from another sewage works to the North Sea through a pipeline instead of discharging it via the rivers of Broadland. These actions together caused an immediate step-fall in the phosphorus load discharged to the River Ant and onwards to Barton Broad. In regard to nitrate reduction, an educational programme of advice to local farmers was instigated to persuade them to adopt a more responsible regime in the use of chemicals.

Thus was born the Barton Broad Clear Water 2000 project. As the new century approached, the UK government created the National Lottery's Millennium Commission and that body invited applications from organisations around the country to seek funding for projects that might be judged as suitable to celebrate the millennium. At the Broads Authority, we were confident that our project would fit this requirement in that it would restore a significant part of one of Europe's most sensitive wetland habitats and would at the same time improve a section of the third largest inland navigation in the UK.

The Authority made a successful bid to the Millennium Commission for a contribution of £1.2 million towards the cost of the project, which would not only restore the Broad and its surroundings to a healthy and attractive state, to the benefit of wildlife and public enjoyment, but would also include the reconstruction of the almost-obliterated Pleasure Island in the centre of the Broad

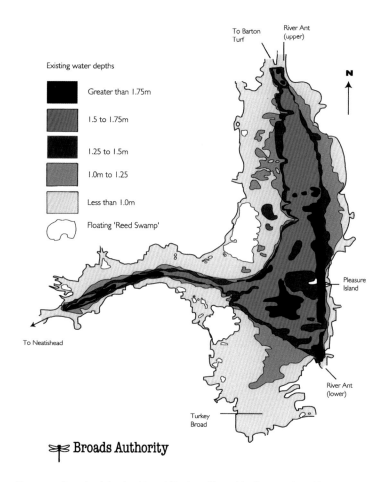

Survey of navigable depths at Barton Broad before restoration. (Broads Authority)

and provide greatly improved public access, both on land and water, throughout this very special area. A further £200,000 for the project came from the Soap and Detergent Industry Association, which had been continuously assisting the Authority with its research over a very long period. The remainder of the £1.5 million budget was provided by the Broads Navigation Committee, which recognised that the removal of the accumulated silt from the Broad would immediately improve navigation, perhaps even re-establishing the conditions that Nelson would have enjoyed.

Barton Broad suction dredger.

Disposal lagoons at Barton Broad.

Advertisements were placed in the Official Journal of the European Union (the OJEU is the central database for European public sector tender notices) and tenders were received from interested contractors. After due process demanded by the National Lottery, a dredging contract was awarded to remove thousands of cubic metres of the oozy mud, which was to be achieved by suction dredging the entire area of the Broad. The output from the dredger (essentially a very large industrial vacuum cleaner mounted on a barge) was to be pumped away through a floating pipeline to the shore and, once out of the water, the pipeline was run through the flooded carr woodland that bounded the lake to terminate eventually at the top of a specially constructed disposal site created on nearby arable land.

The construction of this disposal site was, of course, a prerequisite to the pumping operation. An agreement was reached with a sympathetic local landowner who provided the 21 hectares of land necessary for the settlement of sediment derived from 60 hectares of Broad. A package of financial compensation to this generous landowner was achieved by our negotiating with ADAS, the gov-

ernment agency that approved 'set-a-side' payments. ADAS agreed that because of the overwhelming environmental benefits of the scheme the landowner could receive 'set-a-side' payments even though work was being carried out on his land, which would normally have meant that payments would not have been appropriate. This particular tract of land was chosen because it had a gentle slope down towards the Broad. From an earlier small-scale trial, where sediment from the Broad was deposited into a 20m² lagoon, it had been established that within six months, through natural drying and shrinkage, the sediment would reduce to a third of its former volume. Based on this data the decision was reached that the optimum depth for our disposal lagoons would be 1.5m. The first contractual operation was to carefully bulldoze all of the topsoil from the deposit fields to form linear banks or bunds in order to create a network of lagoons into which the material pumped from the Broad could be placed.

At the other end of the pipeline from the dredger, a mix of 90 per cent water and 10 per cent solids flowed into the lagoons. The pipe, which was up to 1km long, was arranged to discharge this flow

The remote end of the discharge pipe, the other end of which was connected to a dredger out on the Broad.

The disposal fields. (Broads Authority)

into the highest lagoon at the 'top of the hill'. There are in fact no hills in Broadland, but at this site there was just enough gradient to satisfy the engineering requirements of the project. Each lagoon was slowly filled, allowing time for the particulate material to settle out. Once full, the sediment mix then flowed by gravity down to the next lagoon through a series of pipes set into the earth bunds that separated the lagoons. From the lowest lagoon in a particular downhill chain (and there were several of these chains across the area) pipes led through the last bund to allow runoff water to flow into a Terminal Collector Ditch. This ditch was excavated to a gentle gradient back towards the Broad and was lined with a geotextile material that permitted some water to percolate into the local ground structure but prevented the pickup of sediment from the sides of the ditch as the outflow water cascaded down towards the Broad. The ditch was designed to be as long as the local topography allowed, and at three positions along its length renewable filters,

made by inserting bales of hay into the flow, removed any remaining silt. The Environment Agency, with which we had worked in setting up the project, took samples of the now clear water before its final discharge back into the Broad. Before the end of the project all of the light-coloured land in the aerial photograph above had been moulded into further areas of lagoons.

By the time the last lagoon in a particular 'down-hill' chain had been filled, the first lagoon in the chain had already dried and the deposited material had reduced in volume enough to make room for further sediment discharge and, once again, each lagoon was topped up. This process was repeated up and down each chain until dried material in each lagoon almost completely filled the contained void, resulting in a fairly flat surface with the crests of the bunds just peaking through. These bunds, composed of the original top-soil from the fields, were then carefully excavated and the soil was placed on top of half of the area of dried fill material. The exposed half of the dried fill material was then worked back into the grid of voids left by the removal of the topsoil. The whole area was then worked and reworked until the topsoil once again formed the upper-surface layer of the field – a tricky operation but one with which the machine drivers soon became expert. The restored fields now stand testament to the success of this method of operation.

Payment for the dredging work was on a remeasure basis. Comparisons were made between pre- and post-dredged surveys of the Broad at intervals during the five years of the contract. As well as the removal of some 260,000 cubic metres of contaminated silt, also sucked up by the dredger were several outboard motors – and a bicycle. The bicycle, which was identity marked, was returned to a surprised owner, gaining some good publicity for the project. He had lost his bike from the roof of a boat many years before and could not believe that he had got it back, and with surprisingly little damage from its time underwater. On completion of the work, a final post-dredged survey was undertaken. Throughout the period of the works, the project enjoyed good local, regional, national and international press coverage. The Authority provided a step-by-step account to the users of the Broad through its in-house magazine *Broad Sheet*. These navigators did not need to compare the before and after dredging charts, as they could readily appreciate the improved sailing conditions on the Broad and the reappearance of Pleasure Island.

Bio-manipulation, the fourth part of the restoration process, was a science developed internationally by the Broads Authority and by its partner the Environment Agency. Bio-manipulation is centred on the microscopic Daphnia or water flea. A veracious eater of algae if left to its own devices, Daphnia would filter copious volumes of water, removing bacteria as well as algae. A moderate-sized population of Daphnia (say a few tens per litre) could, in ideal conditions, filter an entire lake volume more than once a day. Unfortunately, Daphnia are seriously predated upon by the thousands of small fish

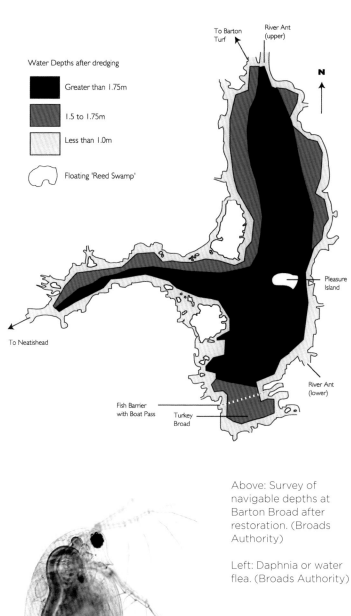

Above: Survey of navigable depths at Barton Broad after restoration. (Broads Authority)

Left: Daphnia or water flea. (Broads Authority)

Aerial view of Barton Broad after restoration. (Broads Authority, Mike Page)

that live in these waters, fish that can remove Daphnia faster than the Daphnia can remove the algae. To overcome this, a balance was 'manipulated' by the introduction into the Broad of larger fish. The introduced pike and perch immediately set about naturally reducing the size of the Daphnia-eating fish populations.

An aerial view of the restored Barton Broad (above) saw it looking pristine in the wider Norfolk landscape. Clearly, the whole area of the Broad would be far too great to attempt to introduce and manage this bio-manipulation process. But, looking more closely, areas of extremely clear water appear to exist in discrete areas of the Broad.

These clear water areas are the controlled areas that have been bio-manipulated by the introduction of additional pike and perch. Once the smaller fish populations had been reduced, the Daphnia

Localised aerial view of part of the Broad after restoration. (Broads Authority; Mike Page)

thrived and ate all the algae, thus clearing the water and allowing light to reach the bottom where submerged plants once again took hold. Damselflies now had somewhere to lay their eggs; the whole ecosystem had been brought back to life.

The controlled areas were separated off from the rest of the Broad using a butyl rubber skirt of sufficient height to allow for the rise and fall of water levels. The skirt was suspended from a 300mm-diameter continuous flotation collar filled with polystyrene. A similar, continuous 300mm-diameter sausage filled with gravel was fixed to the bottom of the skirt to weight it down and seal it to the lake bed. In order to ensure interchange of water between the bio-manipulated areas and the main body of the Broad, we introduced 2m-long replaceable lengths of gauze filter every 50m along the entire length

of the skirt; the gauze was appropriately sized to prevent the entry of small fish.

At Turkey Broad (originally a separate peat digging but now a bay at the southern end of Barton Broad) we had planned to isolate the bay as a bio-manipulated control area, *but* there was an additional consideration. On the edge of this bay, there were several large properties complete with boathouses, the owners of which, literally, had a right of navigation right up to their front doors. So, some means had to be found to keep fish out but, at the same time, allow boats to enter and exit the bay. The solution devised was to build a 'boat pass' into the butyl skirt barrier. I can best describe this crossing point as, essentially, the inverted head of a very large bass sweeping broom through which boats could push but fish would not wish to enter.

The reconstruction of Pleasure Island was achieved by building up the edge of the island using gabions. Gabions are simply wire-mesh baskets filled with natural stone pebbles to form large building blocks. The gabions were topped on their outer edge with a coir (coconut fibre) roll that was planted with indigenous reed-swamp flowers. The contained void within the gabion structure was then filled with selected dredgings and topped off with 300mm of peat material that was planted with local wildflowers.

Included within the Clear Water 2000 project was the improvement of public access in the area of the Broad. Pedestrian access around its periphery was improved by the construction of many boardwalks. These boardwalks not only gave access to pedestrians but also provided access for all. Widened viewing platforms where people might sit and watch, perhaps for the first time ever, the water-based activities on the Broad are also wonderful places for observing the unique wildlife in this corner of Britain. A public trip boat was the other mode of public access provided; the vessel enabled people without a boat of their own to gain an appreciation of the Broad from the water. The drawing submitted as part

Drawing 2
Biomanipulation I - Heron's Carr

Millennium Lottery Fund submission drawing of bio-manipulation proposals for Heron's Carr. (Broads Authority)

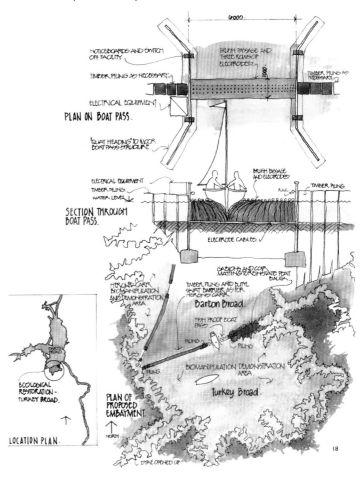

Drawing 3
Biomanipulation 2 - Turkey Broad

Millennium Lottery Fund submission drawing of bio-manipulation proposals for Turkey Broad. (Broads Authority)

of our lottery bid illustrated a typical electric trip boat of the day, and because it needed to be plugged into the mains overnight to recharge the batteries it also included a recharging shed. What was eventually procured went one better. *Ra* is a solar-powered boat, the roof of which consists of solar cells constantly charging the batteries that drive the electric propulsion unit. *Ra* and the accompanying shoreside facilities were also designed and built to provide access for all. Out on the water, those confined to wheelchairs can now share the sense of freedom that such a trip can provide. For those seeking a greater sense of adventure, then a trip on *Electric*

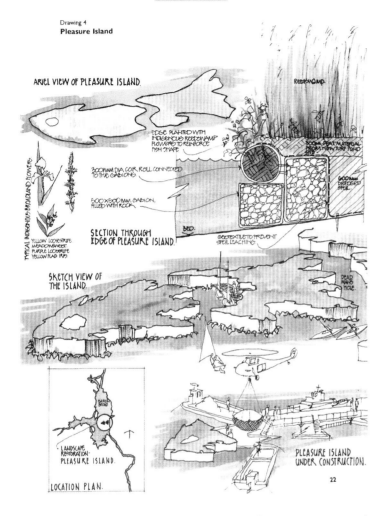

Barton Broad Project

Drawing 4
Pleasure Island

ARIEL VIEW OF PLEASURE ISLAND.

EDGE PLANTED WITH INDIGENOUS REEDSWAMP FLOWERS TO REINFORCE FISH SHAPE

300MM DIA COIR ROLL CONNECTED TO THE GABIONS

500 X 500 MM. GABION, FILLED WITH ROCK

SECTION THROUGH EDGE OF PLEASURE ISLAND.

GEOTEXTILE TO PREVENT SPOIL LEACHING

TYPICAL INDIGENOUS BROADLAND FLOWERS
YELLOW LOOSESTRIFE
MEADOWSWEET
PURPLE LOOSESTRIFE
YELLOW FLAG IRIS

SKETCH VIEW OF THE ISLAND.

LOCATION PLAN.

LANDSCAPE REINFORCED PLEASURE ISLAND.

PLEASURE ISLAND UNDER CONSTRUCTION.

22

Millennium Lottery Fund submission drawing of restoration proposals for Pleasure Island. (Broads Authority)

Eel through the narrow channels fringing the Broad might be the choice, or for the really adventurous (and in true _Swallows and Amazons_ style) a canoe safari can be taken along even narrower connecting waterways around the Broad, where less-disturbed, wildlife is always close at hand.

Public trip boat _Ra_. (Broads Authority)

If the luxury of cruising where everything is done for you is more your style – as Barton Broad is now very much open again for navigation, with sufficient depth even for the traditional Norfolk Wherry – then you might like to cruise on one of these superbly restored vessels that can be hired complete with a skipper and a cook, leaving you to enjoy these tranquil waterways and soak up the restored beauty of this unique cruising area of the UK.

Considerable interest was shown in our project throughout Europe and the Far East. Many delegations visited the site during the work and others have since visited. Exchange visits were made between our team and members of a similar team at the Biesbosch National Park in the Netherlands. Towards the end of the project, the Construction Industry Research and Information Association (CIRIA) had set up a steering group to research 'good practice for inland dredging' and the experience gained at Barton Broad was carried into that group.

So a little more than pure civil engineering, but the time that I spent at the Norfolk Broads was most enjoyable. I hope that this tale of the quiet East Anglian waterways might one day encourage you to go cruising there among the windmills.

Three restored 'pleasure wherrys' sailing on Barton Broad. (Broads Authority)

7

THE SUEZ CANAL

The Suez Canal provides a link between the Mediterranean and the Red Sea and, in a wider context, between the Indian Ocean and the North Atlantic. Shorter journeys with savings in both time and fuel are achieved by ships using the canal. For example, a tanker travelling from Bahrain to Rotterdam would save 7,617km using the canal instead of travelling via South Africa.

For anyone who enjoys ocean travel, to transit this canal just has to be on a 'must-do list'. Gliding through the deserts of Egypt on board a large passenger ship really is a unique experience.

History records that the first linking of the Mediterranean to the Red Sea was achieved in 1887 BC, when Pharaoh Senusret III cut a channel linking the Nile Delta to the Red Sea. His waterway subsequently silted up and was abandoned, but over the centuries channels were reopened on several occasions by each of the leaders who followed him. None of these channels were in existence in 1798 when Napoleon Bonaparte invaded Egypt. However, the engineers, philosophers and other academics that he brought with him (and whose role was to study the country so that France might make maximum use of its resources) knew of these old canals and set about formulating a plan to dig a new one. Unfortunately, errors in calculations made by the leading surveyor in Napoleon's entourage in regard to the relative levels of the Mediterranean Sea and the Red Sea hindered canal development plans for decades.

Meanwhile, in France the views of Henri de Saint-Simon were gaining pre-eminence in French society. Believing that science and technology would better solve most of humanity's problems, he publicly opposed feudalism, militarism and the spiritual guidance provided by the Church. He correctly foresaw the industrialisation of the world and was one of the first to propose that the states of Europe form an association to suppress war, and his ideas were to lay the foundation for a French-built canal. One year after his death, the Saint-Simonian Society was formed and his 'disciples' carried his message far and wide – a message that would have a major

A section of the new north-bound channel of the Suez Canal near Ismailia, Egypt, which opened in 2015.

influence on the intellectual life of nineteenth-century Europe. One of those disciples was Barthélemy-Prosper Enfantin, a French social reformer who thirty-five years after Napoleon had been to Egypt went there himself. Accompanied by a number of engineers, his sole intention was to create a canal across the Isthmus of Suez, believing that such a project would symbolise a marriage between East and West. Enfantin's ideas irritated the Egyptian viceroy, Muhammad Ali, whose permission would be needed to construct such a waterway. Ali wanted the Saint-Simonians' thrown out of Egypt, and they were only just saved from long jail sentences by the intervention of the French consul, Ferdinand de Lesseps. Enfantin would come to regret de Lesseps' help when, twenty years later, he appropriated to himself the Saint-Simonian's ambitious plans to join the two seas with a canal. Certainly not the first person to steal someone else's idea but, in this, de Lesseps was most successful, becoming in later years immortalised as the creator of the canal, the contribution made by Enfantin being largely forgotten. However, in order to deliver the canal, de Lesseps needed the support of the Egyptian ruler.

So, why was a canal desirable? In the early 1800s, communication between Europe and the Far East involved long sea voyages via the Cape of Good Hope. In 1837, the Peninsular Steam Navigation Company secured a government contract for the regular carriage of mail between Falmouth and the Peninsular ports as far as Gibraltar. In 1840 the company won a second contract for the mail service between the UK and Egypt. This new contract was awarded on the condition that within two years the company would establish a line of steamers capable of conveying the mails onwards with a regular service from Egypt to India. Such an undertaking would require the building of larger ships and the establishment of coaling stations, docks and supply and repair facilities at strategic points along the route. In order to raise the considerable capital investment for these developments, the Peninsular Steam Navigation Company (then privately owned by three Victorian gentlemen: Anderson, Wilcox and Bourne) became, in 1840, a limited liability company – the Peninsular and Oriental Steam Navigation Company, incorporated by royal charter. The P&O Company was born!

As contracted, the Indian mail service was inaugurated in the autumn of 1842 by P&O's purpose-built 1,800-ton wooden paddle steamer, *Hindostan*. Additional mail contracts followed and by the end of 1844 P&O was operating a regular mail service extending from Suez to Ceylon, Madras and Calcutta. From 1845, an extension was inaugurated from Ceylon to Penang, Singapore and Hong Kong. In 1849, this route was extended to Shanghai. Steam communication with Australia was inaugurated in 1852 by means of an extension of the line from Singapore. The big drawback to a journey from the UK to any of those destinations was the gap in the through sea voyage between Alexandria and Suez. The only way to travel between these two Egyptian seaport cities was to cross the isthmus on a journey known as the 'overland route'. The first part of that journey was accomplished with goods and passengers being transported from Alexandria to the northern reaches of the River Nile by canal boat. The journey then continued down the Nile to Cairo by river steamer, where passengers disembarked once more for the most arduous part of the journey, a 160km camel trek through the desert to Suez. Approximately 3,000 camels were, apparently, called upon to transport the cargo of a single steamer along this route – a picturesque route, but every passenger and every package was subjected to three separate transfers in travelling from the Mediterranean to the Red Sea. P&O sought to enhance the journey experience for its passengers by providing company-owned rest houses and distracting excursions to Egypt's historic sites. The overland route was, however, not for the faint-hearted.

In 1851, the then viceroy of Egypt, Abbas I, anxious to expand traffic by this important overland route and hence increase revenue to the Egyptian exchequer, entered into negotiations with railway engineer Robert Stephenson for the construction of a railway from Alexandria to Cairo. Begun in 1852, this was the first railway to be built on the African continent. Involving the construction of two bridges over the Nile, this 194km-long line was opened in 1856. The river bridges were built at Benha and Kafr El-Zayat where, prior to completion of the bridge, through traffic was ferried across the river to Cairo. An extension of the railway between Cairo and Suez was opened in 1858 and although the overland route could then be made throughout by rail, it was still only a practicable and commercially viable route for P&O for the carriage of low-volume, high-value cargoes. This was despite taking into account the annual £225,000 subsidy paid by the British government for running the eastern mails. However, until the opening of the Suez Canal in 1869, the railway route was a source of considerable revenue to the Egyptian exchequer, with the wider benefits of the railway being felt by travellers who could now embark on ocean voyages in the reasonable expectation of arriving safely on a predetermined date and even at a given time. It was said you could set your watch by the departures and arrivals of the P&O steamers.

At the time of these developments, Egypt was not a sovereign country; overall power was vested in Constantinople from where the appointment of a viceroy to rule Egypt was made. (Before the Ottoman conquest of Egypt in the sixteenth century the country had already been ruled by a succession of Arab, Kurdish and Turkish dynasties.) Muhammad Ali Pasha commanded the Ottoman Empire army that had driven Napoleon Bonaparte out of Egypt, following which he declared himself ruler of Egypt. After repeated failed attempts by the government in Constantinople to remove him, he was officially recognised in 1805 as pasha and governor of Egypt. He is widely regarded as the founder of modern Egypt because of the dramatic reforms he made to the military and to the economic and cultural life of the country. The dynasty that he established would rule the country until the Egyptian Revolution of 1952. In line with his grand ambitions, he styled himself 'khedive' (viceroy), as did his successors, Ibrahim Pasha, Abbas I and Sa'id I.

In 1835, 30-year-old Ferdinand de Lesseps, an officer in the French diplomatic service, was appointed consul to Egypt. With a passion to build a canal across the isthmus, he nurtured a relationship with Muhammad Ali, a relationship that developed through somewhat unusual circumstances. Ali had a teenage son, Sa'id, who was grossly overweight, and he was determined that his son should lose his excess weight but, despite putting him through rigorous daily exercise with the military, this did not happen. In desperation, he turned for help to an unlikely ally, the French consul. As it turned out, de Lesseps did not have the heart to act as Sa'id's dietician and the two became firm friends, a friendship that later proved key to the development of the Suez Canal.

When Ali died in 1848, power passed to his eldest male relative, his grandson Abbas. Through decree after decree Abbas reversed many of his grandfather's modernisation reforms. He recalled Egyptian students studying abroad and closed schools teaching science at home. He placed restrictions on European merchants, refused to meet with their consuls and made it clear that Egypt was now closed to foreign business. Ferdinand de Lesseps soon realised that there was no point in seeking permission from Abbas to dig a canal. Abbas's six-year reign was brought to an abrupt end when on 13 July 1854 he was murdered in Benha Palace by two of his slaves. As the eldest male descendent of Muhammad Ali, Sa'id Pasha succeeded Abbas.

De Lesseps lost no time in visiting his old friend to congratulate him on his succession. Now nearly 50 years old, he sought from Sa'id an agreement to dig a canal between the Mediterranean and the Red Sea. A concession was granted on 30 November 1854 in which Article 14 importantly stated:

We solemnly declare, for ourselves and our successors, subject to the ratification of his Imperial majesty the Sultan, that the Grand Maritime Canal from Suez to Pelusium will be open always, as a neutral passage, to every commercial vessel crossing from one sea to the other, without any distinction, exclusion, or preference of persons or nationalities.

De Lesseps immediately embarked upon an international campaign to seek support and finance for the construction of the canal. His chosen method was to sell shares in a Universal Maritime Suez Canal Company. When the subscription eventually closed, of the 23,000 persons who had purchased shares, 21,000 were French citizens. From the outset, the British government, led by Lord Palmerston, was very much against the construction of the canal, which was seen as a French enterprise and against the interests of Britain and her empire. Even though de Lesseps recognised the distinction between the reception given to his project by the British government and that accorded him by the British people, he was surprised that not a single share was purchased in Britain. The P&O Company had reason for opposing the building of a canal, believing that once opened, its capital investment in the development of the overland route would be lost and its monopoly on eastern mail and passenger services would soon disappear. They had a large establishment at Suez with distillation plants supplying freshwater, an ice-making facility, stores, repair shops, a coal station, lighters and tenders. At Alexandria, they owned offices, warehouses, docks, coal stocks, coal hulks and many service vessels. Near Cairo and at other locations, they owned farms producing chickens, sheep, vegetables and fruit to supply their steamers at Suez and Alexandria. All of this would become obsolete once ships were built that could take advantage of the canal and provide non-stop voyages between the UK and the Far East.

Work officially began on building the Suez Canal on 25 April 1859, a project that was going to take ten years to complete using mainly Egyptian labour. Throughout the construction period, when

Ferdinand de Lesseps, the French diplomat responsible for the Suez Canal. (National Library of France (BnF), via Wikimedia Commons)

he was not away on his many fundraising expeditions, de Lesseps lived in a house at Ismailia, a town established on the banks of the Suez Canal at roughly its midway point. One of the early tasks to be tackled was the construction of a freshwater canal from the River Nile at Cairo across the desert to Ismailia, from where it would be continued to the north towards Port Said and to the south towards Suez to distribute drinking water to all of the Suez Canal construction sites. When completed throughout in 1863, a celebration was held near Suez to mark the arrival of the first vessel to make the complete journey from near Port Said, not along the Suez Canal but along the freshwater canal. This smaller-scale waterway also irrigated a wide strip of land to which the Suez Canal Company had been given rights and which would rocket in value as it became available for agriculture. (On the death of Sa'id Pasha in 1863, Isma'il Pasha came to power and immediately reneged on the concessions made to the Suez Canal Company in regard to the land irrigated by the Fresh-Water Canal. The crisis was referred during 1864 to the arbitration of Napoleon III, who awarded £3.8 million to the company as compensation for the losses they would incur by the changes to the original grant of concession.).

In the beginning, all excavation of the Suez Canal was undertaken by hand digging, with excavated spoil being moved away from site by traditional means on the backs of camels. The work was carried out by 'pressed' Egyptian workers, operating rather like a National Service organisation, but the regime was described by the British parliament as slavery. De Lesseps soon realised that progress was too slow and he encouraged his engineers to develop machines that could do the work more quickly. Their designs mounted a dozen or so curved steel buckets onto a continuous chain driven by a steam engine and, as a result, efficiency was increased a thousand-fold. Few photographs exist of early construction work, but from illustrations in magazines and journals of the time it is evident that railways or tramways were used to remove excavated spoil from the canal

Hand digging the Suez Canal and removing excavated spoil on the backs of camels. (Suez Canal Authority)

work sites. Once dry excavations were flooded, work continued using floating dredgers, which had a number of cutting buckets running on a continuous chain, similar to land-based dredgers. Dredging practice was further developed by the use of separate floating conveyors that were loaded at one end by the dredger and a conveyor belt then moved excavated material some distance to the canal bank.

As longer lengths of canal were flooded, arrangements had to be made to allow those other 'ships of the desert', camels, to cross the waterway and follow their ancient trails. Simple floating bridges were provided, which were pulled across the canal on ropes by the ferrymen.

Other major elements of the canal infrastructure included the two long breakwaters at Port Said on the Mediterranean coast. The breakwaters trapped sand as it drifted along a coast that would otherwise fill the canal trench that had been cut across the original beach. The completed western breakwater at one time bore at

Floating conveyors designed to move excavated spoil over wide distances. (Author's collection)

Double-tracked canal near Ismailia, Egypt. The new north-bound channel (in the foreground) opened in 2015.

One of twenty cutter-suction dredgers engaged simultaneously on the Suez Canal expansion project, 2015.

its centre a large bronze statue of de Lesseps. Unfortunately, that statue is no longer there, having been blown up by an angry mob in December 1956.

The Suez Canal Authority has published data stating that approximately 1.5 million people worked on the construction of the canal, and of these a staggering 120,000 lost their lives – obviously figures impossible to verify.

As the works progressed, the canal was allowed to fill with water in stages. In March 1862, Mediterranean waters flowed as far as Lake Timsah. Seven years later, the waters reached the Great Bitter Lake and five months after that Red Sea water filled the Small Bitter Lake. Three days later, the two seas were joined together in the Great Bitter Lake. On 17 November 1869, just three months after the joining of the waters, the Suez Canal was officially opened at a glittering ceremony attended by some 6,000 dignitaries, including most of the crowned heads of Europe and leaders from further afield. The guest of honour at the ceremony was Empress Eugenie of France, who arrived on board the ship *Agile* as the British squadron fired a salute. Hundreds of tents were erected for the guests together

with a luxurious palace and an opera house, and more than seventy vessels were made available on Lake Timsah for their enjoyment. Thousands of ordinary people lined the canal bank to watch the first convoy transit the waterway during the inauguration ceremony. The Suez Canal was now open for business. In narrow sections of the canal, ships could not pass one another when both were under power; instead, one had to make fast to the bank while the other moved slowly past.

In 1875, a financial crisis in Egypt forced Isma'il Pasha to sell both his personal canal shares and those of his government. The British government jumped at the opportunity to buy into what was now a successful project, taking all of Egypt's shares for just under £4 million.

Continuous improvements to the canal operation have been made throughout its working life, and in 1887 night-time transits were allowed for the first time. As originally constructed, the channel was approximately 8m deep, a dimension that allowed transit of the largest ships of the day. As ships have increased in size, so too has the depth of the channel. Studies were put in hand in 2010 to determine if the maximum permitted draft of a transiting ship could be increased safely from 20m to 22m, which the Canal Authority stated would cater for 99 per cent of all of the ships in the world. Until 2015 most of the canal was still single-way except for some isolated 'two-way' lengths that had been added over the years to allow ships to pass one another. In 1955, the 10km-long Ballah bypass was opened. In 1980, the 27km-long bypass channel at Port Said was opened, allowing ships direct access to and from the Mediterranean without the need to pass through the port. Also opened in 1980 were the 4km-long Timsah bypass and the 27km-long Difresoir & Kabreet bypass.

In a recent major expansion project completed in 2015, a new 35km length of canal was added, thereby double tracking on either side of Ismailia the original canal and permitting two-way transit of ships for the first time on this length. Provision has also been made for future residential and tourist areas on the Sinai Peninsula side of this new length of canal at Ismailia, to be accessed eventually by a second tunnel under the waterway. Another element of the expansion project was the deepening and widening of the Ballah bypass and the 27km-long channel across the Great Bitter Lake. It was not the aim of this expansion project to make two-way traffic possible over the entire length of the canal but, where this is not yet possible, further passing places will be added at a future date. Shipping will continue to transit the canal in convoys but individual convoys can now be very much longer with passage times greatly reduced. Southbound convoys now average eleven hours instead of the eighteen hours taken before completion of the most recent expansion project. The full extent of the work undertaken at Ismailia shows clearly on recent satellite images published on Google Earth.

The expansion project, which in addition to the excavation of new lengths of canal also included the widening of existing channels to 400m with a depth of 24m and the increasing of the radii of all canal turns to a minimum of 5,000m, was completed within an amazingly short period of just one year. The work was undertaken using a fleet of twenty cutter suction dredgers of different sizes and a number of trailing suction hopper dredgers, claimed to be the largest dredging fleet ever deployed on a single dredging contract. To achieve this timescale, an average of 1 million cubic metres of material was dredged every day, with much of it being pumped ashore through floating pipelines. When at full production, the expansion project employed approximately 1,800 people, either onshore or afloat on one of the dredgers or support vessels. On 6 August 2015, with much pomp and ceremony, Egypt's new addition to the Suez Canal was declared open by Egyptian President Abdul Fattah el-Sisi aboard the historic yacht *El Mahrousa*, which had been present at the original opening of the canal in 1869. Aboard the yacht with him were many foreign dignitaries including French President François Hollande and Russian Prime Minister Dmitry Medvedev.

Linking the Sinai Peninsula with the Nile Valley, there are fourteen busy ferry crossings operating thirty-six ferry boats. There are two fixed road crossings of the Canal: the Al Shahid Ahmed Hamdi Tunnel, located to the north of Suez city; and the Mubarak Peace Bridge, a cable-stayed design, built between 1996 and 2001 and located in southern Port Sa'id. The bridge has a navigation span of 404m and its clearance above high-water level of 70m means that the largest ships in the world can transit the waterway. The El Ferdan Railway Bridge is the largest railway swing bridge in the world. When it was closed across the canal for trains to cross, the main span was 340m and each of the side spans were 150m. Completed in 2001, it replaced an earlier bridge of similar design. The replacement bridge was opened by President Mubarak, but it is no longer in use because the new length of canal completed in 2016, providing a dual waterway, has severed the rail connection into the Sinai.

Due to political upheavals centred on Egypt, the operation of the canal has not been continuous throughout its history and the

Mubarak Peace Bridge.

canal has been completely closed to shipping on several occasions, the longest closure being a period of eight years from June 1967 to June 1975.

In 1882, during the administration of Khedive Tewfiq, the British government argued that Egypt was descending into anarchy, which was threatening the operation of the Suez Canal, and it sought international support for an invasion of Egypt in order to protect the waterway. In August 1882, as neither the Ottoman sultan nor any of the European governments offered support, Britain acted alone and invaded Egypt. Within two months, British forces, having defeated the Egyptian army at the **Battle of Tel-el-Kebir**, had captured the Suez Canal, leading to its closure for five days. The British prime minister, William Gladstone, tried to withdraw the British forces immediately, but there was no effective Egyptian government left to maintain order and so began seventy-two years of British occupation of the country. Through this occupation, Britain had acquired physical control of the Suez Canal, although France, which had dominated the canal up until that time, still controlled the majority of shares in the Suez Canal Company. In the hope of weakening British control, France attempted to sway European opinion in favour of 'internationalising' the canal. The two powers reached a compromise through the Convention of Constantinople, a treaty signed on 29 October 1888 by the UK, Germany, Austro-Hungary, Spain, France, Italy, the Netherlands and the Russian and Ottoman

El Ferdan Railway Bridge.

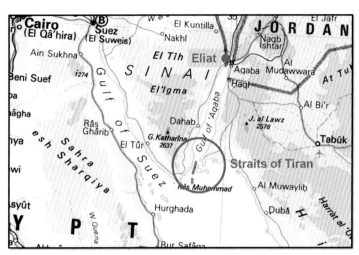

Map of the Straits of Tiran.

empires. Article 1 of the treaty 'guaranteed rite of passage of all ships in time of peace and in time of war'; Article 10 allowed the Ruler of Egypt to take measures for 'the defence of Egypt and the maintenance of public order'.

At the beginning of the **First World War** and, in accordance with the treaty, Egypt declared that the canal would be open to ships of all nations. Britain responded by declaring Egypt a British Protectorate and barred canal access to ships of her enemies, citing the security of the canal as the reason for its action. There were many attacks on the canal during the First World War, although no records exist of any long-term closure of the waterway. One of the most significant attacks was by Turkish forces from across the Sinai, which were repelled by British and French warships shelling from positions within the canal.

Again, in the **Second World War**, Britain invoked Article 10 for the controlling measures taken to ban enemy ships from using the canal. The canal was closed due to enemy action for a total of seventy-six days during the Second World War.

In 1954 Gamal Abdel Nasser came to power in Egypt and soon afterwards Britain signed an agreement ending its seventy-two-year presence in the country. In 1955, Nasser began to import arms from the Soviet Bloc to build up his arsenal for confrontation with Israel. Escalation continued with an Egyptian blockade of the Straits of Tiran, preventing Israel's access to her southern port of Eilat via the Gulf of Aqaba. At the same time, Egypt closed the Suez Canal to Israeli shipping. The United States and Britain had earlier promised to give aid to Egypt towards the construction of the Aswan High Dam on the River Nile but the offer was later withdrawn. In July 1956, Nasser announced to a huge cheering crowd in Alexandria and to the rest of the world via Cairo Radio that he was nationalising the Suez Canal Company and creating an Egyptian Canal Authority to manage the canal and to use the funds raised from its operation to pay for the dam. Immediately afterwards, Egypt went on to sign a tripartite agreement with Syria and Jordan placing Nasser in command of all three armies.

Britain, France and Israel secretly agreed on a plan whereby Israel would land paratroops near the canal and send its armour across the Sinai Desert facing the Egyptian army on the other side of the canal. The British and French would then call for both sides to draw back from the Suez Canal Zone, fully expecting the Egyptians to refuse. At that point in the plan, British and French troops would be deployed in order to protect the canal. Israel launched its attack across the Sinai on 29 October 1956 and, as expected, the Egyptians ignored the Anglo-French ultimatum to withdraw back from the Suez Canal. On 30 October the United States sponsored a Security Council resolution calling for an immediate Israeli withdrawal, but Britain and France vetoed it. The Israeli Defence Force continued to pour across the desert, capturing virtually the entire Sinai within five days. On 1 November, as Egyptian forces had still not moved back from the canal, the Anglo-French plan to 'protect' the canal was put into action; the RAF launched Operation Musketeer by bombing Egyptian airfields. On 5 November, British and French paratroops landed near Port Said as commandoes were landed from amphibious ships. This conflict, afterwards known as the **Suez Crisis**, was a major operation for British forces. Five British aircraft carriers were involved in the operation including HMS *Eagle* (R05), *Bulwark* (R08) and *Albion* (R07). Egypt's response to this invasion was to deny use of the canal to Britain and France and, of course, to the rest of the world, by sinking forty ships in the waterway, blocking all passage. The United Nations sought to resolve the conflict and pressured the two European powers to back down.

Although British forces performed remarkably well, in the eyes of most of the world the British government was humiliated and Prime Minister Sir Anthony Eden's political career was destroyed. While the British and French governments had failed to accomplish their goals, the Israelis were initially satisfied at having captured the Sinai Peninsula in an operation that took just 100 hours. Nasser said that the canal would stay blocked as long as a single foreign soldier remained on Egyptian soil. Much of the rest of the world shunned Britain and France for their actions in the crisis and, after they had backed down, a United Nations salvage team moved in to clear the canal, which remained closed to international shipping for 159 days between November 1956 and April 1957. Once cleared, control of the canal was given back to the Egyptian government with the proviso that all vessels of all nations had free passage through it. Closure

and its aftermath had demonstrated the strategic importance of the canal to Europe, it being the main passageway for getting oil to the Continent, which was soon running in short supply. In March the following year, under pressure from the USA, Israel was forced to withdraw from the Sinai without obtaining any concessions from Egypt. A United Nations Emergency Force was deployed to police the area but tensions continued to rise.

Egyptian President Gamal Abdel Nasser was keen to unite the Arab world and spoke of 'the destruction of Israel'. In May 1967, he demanded the removal of the United Nations peacekeeping force from the Sinai and, at the same time, once again closed off the Straits of Tiran to Israeli shipping. On 30 May 1967, he signed a defence pact with Jordan and invited the Iraqi army to begin deploying troops and armoured units in Jordan. Iraqi troops were later reinforced by an Egyptian contingent. These three actions, together with Egyptian troop deployment in the Sinai, were seen by Israel as a clear sign of Arab preparation for all-out war.

So on 5 June 1967, Israel launched Operation Focus, a large-scale surprise air strike on Egypt, claiming the element of surprise was the only way it could stand any chance of defending itself against the increasing threat from neighbouring states. This operation was the opening of the **Six-Day War**. The Israeli army sped through the Sinai Peninsula driving Egyptian tank forces twice their size right back to the Suez Canal, where just 10 per cent managed to make it back across the waterway. Israel dug in alongside the canal in an occupation of Sinai that was to last for many years. The Israeli navy reopened the Straits of Tiran at Sharm El Sheikh, allowing access once more to their southern port of Eilat. The Suez Canal remained closed.

Arab revenge came six years later with the **Yom Kippur War**. Israel was, at the time, led by Golda Meir and Egypt by Anwar Sadat. On Saturday 6 October 1973, the combined forces of Egypt and Syria launched a surprise attack across the canal, knowing that the Israeli military would be participating in religious celebrations asso-

Left: British Prime Minister Sir Anthony Eden (1955–57). (William Little, via Wikimedia Commons)
Right: Egyptian ruler Gamal Abdel Nasser (1954–70). (Wikimedia Commons)

ciated with Yom Kippur. Just 500 Israeli soldiers in the canal area faced 80,000 Arab soldiers and were swiftly overwhelmed. Within two days, the combined Arab force had moved up to 25km into Sinai and by the end of 7 October the military signs were ominous for Israel. However, the next day, Israeli forces, benefitting from intelligence provided by the USA, achieved a remarkable military recovery. Counter-attacking Egyptian and Syrian forces in the Sinai, they crossed the Suez Canal south of Ismailia from where they advanced to within 105km of the Egyptian capital using the Suez–Cairo road. On 24 October, a ceasefire was arranged by the United Nations, once again sending its own peacekeeping force to the highly volatile Sinai Peninsula. Between January and March 1974 Israeli and Egyptian forces disengaged along the Suez Canal as American Secretary of State Dr Henry Kissinger, acting as a peace broker between them, arranged the signing of an interim agreement in September 1975. Each party declared its willingness to settle differences by peaceful means rather than by military force. This was to lead to American-sponsored talks at Camp David, which

followed the 1977 'Sadat Initiative'. To some Arabs, Anwar Sadat had betrayed the Arab cause and this was to cost him his life when, in 1981, he was assassinated by Muslim fundamentalists.

Following these two wars, and after the removal of many temporary and semi-permanent blockages to the canal (as well as much unexploded ordnance) and the completion of many years of deferred maintenance, the canal was eventually reopened on 5 June 1975 at a ceremony held at Port Said. The Suez Canal had been closed for eight years from 5 June 1967 to 5 June 1975. Not a good record then of continuous operation for an international waterway, particularly when compared with the Panama Canal, which had remained operational on every day of its 100-year history, except for a couple of years ago when it closed for forty-eight hours due to exceptionally high water levels due to a period of storms in Gatun Lake catchment.

The **Arab Spring of 2010** saw civil strife once again return to Egypt. The unrest caused a collapse of the previously thriving tourist industry and has directly affected the use of the Suez Canal. Containership transits have, in recent years, been at an all-time low and shipping companies have complained about an additional 2.5 per cent rise in transit fees imposed by the Egyptian government in order to offset the slump in tourism income. When I transited the canal in 2014, and again in 2017, all seemed tranquil, but you don't have to look far beneath the surface to become aware of the problems that lie there. All along the canal there was evidence of military activity. Soldiers could be seen on guard at regular intervals all along the Sinai bank. Floating bridge pontoons stood ready at numerous locations and ferry crossings appeared to be loaded with military personnel.

Who knows what the future holds for the canal in the volatile Middle East. On my recent transits, Suez city did look beautiful in the early morning sunshine, so let us hope that the Suez Canal has a brighter future than its past might suggest for it.

Military floating bridge pontoons stand ready for action at numerous locations along the canal.

8

THE SEVEN WONDERS OF BRITAIN'S INLAND WATERWAYS

Most of the structures featured in this chapter were built at least 100 years before the construction of the Panama and Suez canals, during Britain's Industrial Revolution. Now, more than 200 years old, these unique examples of the work of the civil engineer can still be accessed either by boat or on foot. Many of the engineering principles used to build these structures were further developed and used to build the Panama and other worldwide canals.

Robert Aickman – author, canal historian and co-founder of Britain's Inland Waterways Association – compiled the original listing of 'Seven Wonders of the Waterways' over half a century ago. Aickman's list, published in his book *Know Your Waterways*, comprised the Pontcysyllte Aqueduct on the Llangollen Canal; Standedge Tunnel on the Huddersfield Narrow Canal; the Bingley Five Rise Locks on the Leeds & Liverpool Canal; Burnley Embankment, also on the Leeds & Liverpool Canal; the Caen Hill Locks on the Kennet & Avon Canal; the Anderton Boat Lift, linking the River Weaver to the Trent & Mersey Canal; and the Barton Swing Aqueduct, which carries the Bridgewater Canal over the Manchester Ship Canal.

Half of the population of the UK lives within 8km of a canal or navigable river and, today, an estimated 11 million people use the canal towpaths annually as part of their everyday life – as a short-cut to work, walking the dog, cycling to the countryside or simply taking time out to watch the boats go by.

The Pontcysyllte Aqueduct

The Pontcysyllte Aqueduct, located on the Llangollen Canal in North Wales, was designed by civil engineer Thomas Telford. Built between 1795 and 1805 by the Ellesmere Canal Company, the aqueduct was part of an ambitious plan to join Ellesmere Port on the River Mersey, opposite Liverpool, to Shrewsbury on the River

Telford's Pontcysyllte Aqueduct, Llangollen Canal. (British Waterways)

9" Upstand above water level

Minimal freeboard of the navigation channel on the Pontcysyllte Aqueduct.

tapered them towards the top in order to keep the weight as light as possible so as not to overload the river-valley soils into which their foundations sit. The mortar used to join the masonry blocks together is made of oxen blood, lime and water. Some years ago my youngest daughter described the experience of navigating across this structure as like 'flying a boat'. In July 2009, UNESCO granted Thomas Telford's masterpiece the designation of a World Heritage Site. It now, rightly, enjoys the same status and protection as does the Taj Mahal and the Acropolis. One other function performed by Telford's structure, as well as greatly adding to the enjoyment of travel on this popular recreational waterway, is that the aqueduct, and indeed the whole of the Llangollen Canal, conveys raw water destined for human consumption from the River Dee at Llangollen to storage and treatment reservoirs at Hurleston in Cheshire.

Standedge Tunnel

Standedge Tunnel is the second structure on Aickman's list. The tunnel is located on the Huddersfield Narrow Canal, which runs for 32km through seventy-four locks between Huddersfield in West Yorkshire and Ashton-under-Lyne in Greater Manchester. At its southern end this waterway connects with the Ashton Canal and at its northern end with the Huddersfield Broad Canal. Standedge Tunnel is the longest, deepest and highest canal tunnel in Britain. It is just over 5km long, runs 194m underground and, at a level of 197m above sea level, the canal through the tunnel is the highest navigable waterway in Britain.

The setting out of the tunnel in 1794 by civil engineer Benjamin Outram was not an easy exercise. A straight line was first laid out across the hills and moors above and a calculation was then made at specific locations to determine just how deep below the ground the canal would need to be. At these locations vertical shafts were

Severn. Much of this route was never completed and, in the end, this fantastic aqueduct only served a basin at the remote village of Ruabon and a short 'branch canal' to the Welsh market town of Llangollen. The iron structure is supported on eighteen stone piers measuring 38m from ground level to the underside of the ironwork. The canal itself runs within a 305m-long iron trough that is 4m wide and 1.6m deep. A towing path, originally for the boat-towing horses, is cantilevered out from one side of the trough providing an exciting walk with spectacular views across the Dee Valley.

The footpath has a safety fence to one side but, on the other side of the canal, there is nothing more than 23cm of exposed iron trough to help prevent boats from a 41m fall to the fields below. Holding one-third of a million gallons of water, the aqueduct takes two hours to drain into the River Dee below from the time that a drain, built into the bottom of the trough, is activated – a process that is undertaken regularly for inspection and maintenance purposes. Pontcysyllte is the highest and longest aqueduct in Britain. The nineteen iron arches each span 14m between the masonry piers. Telford designed these slender piers as hollow structures and

Above: Standedge Tunnel Visitor Centre, Huddersfield Narrow Canal. (Tim Green from Bradford, via Wikimedia Commons)

Right: Standedge Canal tunnel (right-hand side) and three railway tunnels run under Saddleworth Moor in the Pennines.

sunk to the required depth and then the tunnel was dug away from the bottom of the shaft in two directions. From inside the tunnel, although it can be seen that alignment was not always perfect, the resulting bore was a considerable achievement in its day. Early surveys suggested that the rock through which the tunnel would be cut was composed, principally, of strong shale, which would not have presented any serious difficulties; however, this proved not to be the case and much gunpowder had to be used to blast through solid gritstone rock. Large volumes of water entered the workings and brick linings were needed at many more locations than originally thought necessary, thus pushing up construction costs. Thomas Telford completed the tunnel in 1811, six years after Outram's sudden death and thirteen years after Outram had successfully opened the rest of the canal. The tunnel was used by an average of forty boats daily until the mid 1840s, when the number of transits began to fall away.

The Huddersfield Narrow Canal Company was taken over in 1846 by the local railway company, and its civil engineers set about building a railway tunnel under Saddleworth Moor parallel to the canal tunnel. The canal tunnel proved most useful in assisting with the construction of the 1849 single-track railway tunnel, as no vertical shafts were required in its construction. A series of horizontal cross passages were cut linking the railway excavations across to the canal tunnel and, through these passages, all of the excavated spoil from the railway tunnel was removed by canal barge. A second single-track railway tunnel was opened in 1871 and a further double-track tunnel was completed in 1894; it is this tunnel that today carries all trans-Pennine rail services. Notwithstanding severe

Drilling the canal tunnel roof prior to the insertion of anchor bolts. (British Waterways)

competition from the railway, the canal continued to operate until the passage of the last commercial boat in 1921, after which the tunnel soon fell into disrepair, with several large rock falls inside. It was not until 1943, however, that it was officially closed and sealed off by large iron gates.

As part of the canal restoration movement in Britain a decision was made to restore the derelict Huddersfield Narrow Canal as a leisure waterway. At the time of the restoration of Standedge Tunnel the original single-track railway tunnels lay unused, having been closed some forty years earlier by Transport Minister Dr Beeching, as part of his massive closure of much of Britain's railway infrastructure. Now it was the turn of the old railway tunnels to repay the canal tunnel for its help during their original construction; tarmacadam was laid down in each of them, forming roadways to provide access to the canal-tunnel restoration works, the canal itself being accessed via the many cross passages originally constructed to build the railway tunnels. During restoration I visited the workings on a number of occasions; it was an uncanny sensation, as a train roared through the nearby live rail tunnel, to hear it apparently bearing down on the workforce in the dimly lit canal tunnel as the deafening noise of its passing transferred from one cross passage to the next. Several unlined lengths of the canal tunnel were found to be unstable and were strengthened using rock anchor bolts drilled into the roof and used in combination with a sprayed concrete lining. The restored tunnel together with the remainder of the canal was reopened in May 2001.

Due to the length, the narrow width and the shortage of ventilation shafts within the tunnel, British Waterways (the tunnel operator) originally deemed it not safe to allow boats to be navigated through under the power of their own (usually diesel) engines, and instead provided electric tugs to pull boats through. A change in arrangements now permits owners to navigate through the tunnel under their own power, but with a local guide (or pilot) being provided for each transiting boat. But you don't need your own boat to experience a subterranean cruise under the Pennines – a 'voyage with a difference' can be made through the tunnel in specially designed passenger boats departing from the visitor centre located at its northern entrance.

The Bingley Five Rise Locks

The Bingley Five Rise Locks are located on the Leeds & Liverpool Canal, a canal that makes an end-on connection with the Aire & Calder Navigation offering a coast-to-coast route from the Irish Sea at Liverpool right across the north of England to Hull and the North Sea, passing through the beautiful northern countryside and fascinating towns and villages en route. The Bingley Locks operate as a 'staircase' flight in which the lower gate of one lock chamber forms the upper gate of the next – exactly the same design as used 140 years later on the Panama Canal. Designed by the Leeds & Liverpool Canal's chief civil engineer, John Longbotham, the locks

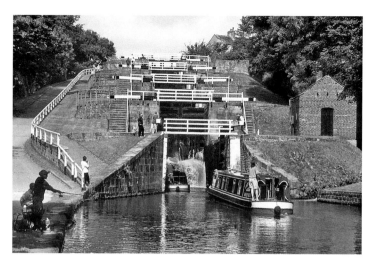

Bingley Five Rise Locks on the Leeds & Liverpool Canal.

were completed in 1774. On the opening day, thousands gathered to watch the first vessels make the 18m ascent. Now, almost 250 years later, the flight is still in daily use providing access to 26km of lock-free cruising through the glorious Yorkshire Dales. The Bingley Five Rise Locks, like all staircase locks, must be 'set' before beginning passage. For a journey upwards, the bottom lock must be empty, with all of the others full of water. It can take up to ninety minutes for a boat to work through this flight, being either lifted or lowered by 18m, but navigators should not be in a rush on this leisure waterway. A holiday on Britain's canals is often described as 'the fastest way to slow down'. At 204km long, the Leeds & Liverpool Canal is the longest canal in Britain; yet it is, today, one of the least busy of the three trans-Pennine leisure waterways.

Burnley Embankment

The impressive Burnley Embankment, also located on the Leeds & Liverpool Canal, is another of Aickman's waterway wonders. At nearly 1.5km long (almost 1 mile, and known locally as the 'straight mile'), it was designed by civil engineer Robert Whitworth and was built between 1796 and 1801. The embankment carries the Leeds & Liverpool Canal across the valleys of the rivers Brun and Calder without the need at each location to lock down on one side of the valley and up again on the other.

Caen Hill Locks

Further south now, to the Kennet & Avon Canal and the Caen Hill Locks, which, when completed in 1810, provided an inland navigation for the carriage of freight between the Port of Bristol and the Port of London via three local rivers (the Bristol Avon, the Kennet and the Thames); a route avoiding French privateers who were, at that time, attacking British shipping along the south coast of England. The flight of sixteen locks at Caen Hill is located in the middle of the Devizes flight of twenty-nine locks in Wiltshire. The flight is not the longest in the UK (that honour belongs to the thirty-six locks at Tardebigge on the Worcester and Birmingham Canal), but Caen Hill is visually far more impressive because its sixteen locks fall in a straight line and are built very close together. They are, though, individual locks with gates at each end and not part of a staircase where gates are shared between adjacent chambers, as at the Bingley Five Rise Locks and on the Panama Canal. Caen Hill Locks were the last element of the 134km-long Kennet & Avon Canal to be built by civil engineer John Rennie.

As an aside, after completing his work on the Kennet & Avon Canal, Rennie went on to build a 'Bridge Across the Atlantic', and that story has been recorded in a book of that name, written by Wallace Reyburn. A bridge right the way across the Atlantic Ocean does of course not exist, and never will. Reyburn's book tells the story of Rennie's London Bridge, which he built across the Thames in 1831 and which, when replaced by the current London Bridge, was offered for sale. It was subsequently purchased by a group of

Caen Hill Locks on the Kennet & Avon Canal. (British Waterways)

American businessmen who took it to pieces stone by stone and transported it across the Atlantic, hence the name Bridge Across the Atlantic. These businessmen then reconstructed the bridge as a centrepiece for the new leisure resort of Lake Havasu City in Arizona. According to some American friends of mine, there just might be some truth in the rumour circulating at the time that those American businessmen thought that they were bidding for Tower Bridge, which is popularly known in the United States as the iconic 'London Bridge'. Who knows if there is any truth in that rumour?

Back to John Rennie's earlier work. The Kennet & Avon Canal operated without the Caen Hill Locks for a short while using a gas-lit tramway to convey goods up and down the hill between waiting boats. Lit to provide for a twenty-four-hour operation, it is widely believed that this was the first gas-lit 'highway' in Britain. When I managed the restoration of the flight a dozen or so years ago, we discovered remnants of that old gas lighting system buried within the towpath.

Because the locks were built so close together, each chamber is provided with an adjacent side reservoir. These reservoirs supply operational water to each of the sixteen locks climbing the hillside. On the average run of canal, locks are spaced approximately 1–1½km apart so there is sufficient water in the pound between the locks for their operation. At Caen Hill there would be insufficient water between them for operation, hence the provision of a reservoir or side pound at each lock.

Before leaving this waterway, I should mention that at its extreme western end, in the Bristol City Docks, the restored steamship *Great Britain* is open as a multi-award-winning museum. The SS *Great Britain* story is covered in Chapter 5.

The Anderton Boat Lift

By the end of the seventeenth century a major salt-mining industry had developed around the Cheshire 'salt towns' of Northwich, Middlewich, Nantwich and Winsford. The completion of the River Weaver Navigation in 1734 provided a navigable route for transportation of salt from the mines to the River Mersey. The opening of the Trent and Mersey Canal in 1777 provided a second transport route that ran close to the Weaver Navigation for part of its length, but extended further south to the coal-mining and pottery industries around Stoke-on-Trent. Rather than competing with one another, the owners of these two independent waterways decided that it would be far more profitable to work together. In 1793 a basin was excavated on the north bank of the Weaver at Anderton, which took the river to the foot of the embankment, supporting the canal some 50ft above. Facilities were built at this location for the trans-shipment of goods between the two waterways, including the provision of two salt chutes. These facilities were extended in 1801 and again in 1831, and then, later in the nineteenth century, a lift was built to move fully loaded boats from one waterway to the other.

The Anderton Boat Lift, which opened in 1875, was designed by civil engineer Edwin Clark. It consisted of two counterbalancing, open-topped, water tanks or caissons, each large enough to take a fully loaded barge or a pair of traditional narrow boats. Hydraulic rams located beneath the tanks provided the vertical motion; the main force powering this movement was gravity. The descending caisson was simply loaded with an extra 15cm of water and, as this heavier caisson descended, the compressed water in the cylinder beneath it passed through a connecting pipe into the adjoining cylinder where its ram, linked to the second caisson, pushed that caisson up to the level of the canal. By 1904, these hydraulic cylinders and piston rams were worn out and so the operators asked their engineer, Colonel J.A. Saner, to investigate an alternative method of operation.

Anderton Boat Lift, Trent & Mersey Canal. (British Waterways)

Colonel Saner's solution was to replace the hydraulic mechanism with electric motors and a system of counterweights and overhead pulleys. This arrangement had the additional advantage of allowing each caisson to operate independently of the other. Although involving many more moving parts than the hydraulic system, at least all of these would be above ground and easily accessed for maintenance. The original supporting superstructure had to be greatly strengthened because the addition of the counterweights to balance the movement of the caissons doubled the weight the superstructure had to carry. This strengthening was provided by ten steel A-frames, five on each side of the lift. These frames supported the machinery deck 18m above, upon which the electric motors, drive shafts and cast-iron headgear pulleys were mounted. Wire ropes attached to each side of each caisson passed over the pulleys to thirty-six cast-iron counterweights, eighteen for each caisson. Each counterweight weighed 14 tons, so that eighteen exactly balanced the 252-ton weight of a loaded caisson. The electric motors were only required to overcome friction between the pulleys and their bearings. Conversion work was carried out between 1906 and the reopening date of 29 July 1908. Seventy-five years later, in 1983, problems with the mechanism caused the lift to once again 'permanently' close.

During the 1990s, British Waterways carried out preliminary investigations prior to launching a bid for restoration funding. At first it was announced that the intention was to restore the lift to electrical operation, but in 1997, after consultation with English Heritage, it was decided to restore the lift to hydraulic operation as originally designed. A partnership consisting of the Waterways Trust, the Inland Waterways Association, the Anderton Boat Lift Trust, the Friends of Anderton Boat Lift, the Association of Waterway Cruising Clubs, British Waterways and the Trent and Mersey Canal Society was put together to raise the required £7 million for the work. The Heritage Lottery Fund agreed to contribute £3.3 million, and more than 2,000 private individuals also contrib-

uted to the scheme. Restoration started in 2000 and the lift was once again reopened to boat traffic in March 2002. Although a modified version of the original hydraulic system was reinstated, the 1908 external A-frames and the (now redundant) pulley system on the machinery deck have all been retained in a non-operational role, so as to contribute to the telling of the story of the life of the structure.

Barton Swing Aqueduct

To the north-east of Anderton, the Bridgewater Canal is carried across the Manchester Ship Canal on the Barton Swing Aqueduct.

Originally located at this site was a three-arched sandstone aqueduct, built in 1761 for the Duke of Bridgewater by civil engineer James Brindley. Brindley's structure carried the Bridgewater Canal over the River Irwell 11m below. The Bridgewater Canal, which needed no locks, was constructed to convey coal from the duke's coal mines at Worsley into the centre of Manchester. No aqueduct on this scale had previously been constructed in England and critics thought it would never hold water.

The River Irwell was at that time part of the Mersey and Irwell Navigation. In 1885, both the Irwell Navigation and the Bridgewater Canal were bought by the Manchester Ship Canal Company. Much of the line of the River Irwell was then incorporated into the design for the Manchester Ship Canal and this necessitated the demolition of Brindley's aqueduct to enable much larger vessels to pass beneath the Bridgewater Canal. The sandstone arches were replaced by the Barton Swing Aqueduct, which was designed by civil engineer Edward Leader Williams. Opened in 1893, this unique structure essentially consists of a 71m-long by 5m-wide steel tank, literally containing a length of the Bridgewater Canal. With gates closed at each end by hydraulic rams, 800 tons of water is trapped within the tank, which is then swung round on its pivot situated on an

Barton Swing Aqueduct, Bridgewater Canal. (British Waterways)

Falkirk Wheel and visitor centre, Forth & Clyde and Union canals. (British Waterways)

island in the middle of the Ship Canal, thus clearing the way for ocean-going ships to make passage from the Irish Sea to the centre of Manchester. The total weight of the swinging span, including the contained water, is 1,450 tons. Still in use today, it swings occasionally for passenger ships and other visiting craft transiting the Ship Canal – a far cry from the numerous openings per day needed for the operation of the once extensive and very busy Manchester Docks, now renamed Salford Quays. The whole area has been remodelled as a leisure and tourist facility, bringing together a mix of culture and retail. The Lowry Gallery is located right on this continually evolving waterfront destination and a new tram service links the area to the centre of Manchester.

Seven Wonders of the Inland Waterways for the Twenty-first Century

In 2002, British Waterways (the government department that managed Britain's canal network at that time) conducted a poll of those interested in the inland waterways, asking them to choose the 'Seven Wonders of the Inland Waterways for the 21st Century'.

Those chosen included five already on Aickman's List, but there were two new entries: Sapperton Tunnel on the Thames & Severn Canal and the Falkirk Wheel in Central Scotland.

The 220km canal network in Scotland was built between 1768 and 1822. It consists of the Caledonian, Crinan, Forth & Clyde and Union canals. Although small in number, they are some of the most famous and historic waterways in Great Britain. The Caledonian Canal between Fort William and Inverness, for example, makes use of the whole of the length of Loch Ness, so a cruise along that waterway might reveal a monster for which you did not bargain.

The Falkirk Wheel

The Falkirk Wheel, which links the Forth & Clyde Canal to the Union Canal in Central Scotland, is the odd one out in this chapter, as it was not built 100 years before Panama or Suez but was built during the twenty-first century as part of a restored waterway route right across Scotland from the Atlantic Ocean to the North Sea.

In recent years, British Waterways has reversed years of neglect and has revitalised Scotland's canals, transforming them into an important national asset. More and more Scots and visitors to Scotland are discovering that canals are the perfect way to explore Scotland's wonderful countryside. Part of this revitalisation included the construction of the Falkirk Wheel, which was designed to reconnect the Forth & Clyde Canal to the Union Canal, the eastern terminus of which is right in the centre of Edinburgh.

An exceptional feat of modern engineering, the Falkirk Wheel is already recognised as an inspirational sculpture for the twenty-first century. This elegant mechanical marvel, the only rotating ship lift in the world, was opened on 24 May 2002 by Her Majesty Queen Elizabeth II as part of her Golden Jubilee celebrations. Sad to say, the royal opening was delayed by a month due to extensive damage of the wheel's electrical systems – the work of local vandals.

The Falkirk Wheel is sited in a natural open amphitheatre at Rough Castle near Falkirk, where visitors can enjoy the 'Falkirk Wheel Experience' by cruising on special passenger boats and by visiting the distinctive new visitor centre located at low level beneath the wheel. The centre provides a sensational view of the rotating mechanism through its glass roof, and a one-hour boat trip takes the visitor on a 'rotating' journey to the top of the wheel and back again. Extended boat trips along the two canals are also available. These canals were previously connected by a flight of eleven locks, which, by the 1930s, had fallen into disuse (they were filled in and the land built upon). The plan to regenerate the canals of central Scotland and to reconnect Glasgow with Edinburgh by water was led by British Waterways, with support and funding from the Scottish Enterprise Network, the European Regional Development Fund, the Millennium Commission and seven local authorities. A decision had been made very early on in the regeneration planning process to create a dramatic twenty-first-century landmark structure to reconnect the canals.

The mechanism for keeping the boat-carrying tanks upright as the Falkirk Wheel turns.

The difference in the levels of the two canals at the Falkirk Wheel is 24m. However, the Union Canal, which runs on one level all the way from Edinburgh, is at a higher level than the top of the Falkirk Wheel and so boats have to pass down through two new locks to descend from the canal onto the wheel's approach aqueduct. The reconstructed end of the Union Canal had to be lowered in order to pass under the historically important Roman Antonine Wall and also under the main Edinburgh–Glasgow railway line, both of which cross the boundary of the site.

The mechanism that ensures that the tanks containing the boats remain upright whilst the wheel rotates is quite simple. The large gear wheel 'A' is a fixed stationary 'sun' gear attached to the supporting structure. As the wheel rotates in a clockwise direction, the idle 'planet' gear meshes with the fixed 'sun' gear as it rides around its periphery, causing it to rotate in a clockwise direction pushing gear wheel 'B', which is connected to the tank containing the boats, in the opposite direction to the rotation of the wheel, thereby keeping the tank upright at all times. Irrespective of the number and size of

Coates Portal of Sapperton Tunnel, Cotswold Canals. (Karen Shaw, Cotswold Canals Trust)

vessels in either tank, the wheel is perfectly balanced at all times, obeying Archimedes' principle; as the designed rotational speed is comparatively slow, the power needed to turn the wheel, just to overcome friction, is around 5–12 kilowatts or about the same as that required to boil three or four kettles of water.

From twenty-first-century technology back to eighteenth-century-technology and the second of the new entries in the 2002 'Seven Wonders of the Inland Waterways' list – Sapperton Tunnel.

Sapperton Tunnel

Sapperton Tunnel is near to the halfway point along the Cotswold Canals. Cotswold Canals is the new name for the waterways that were originally built as the Stroudwater Navigation and the Thames & Severn Canal, which connected together with an end-on junction at the town of Stroud in Gloucestershire. A total of fifty-seven locks were required to lift vessels from the River Severn up and over the Cotswold

Hills and down again to the River Thames. Sapperton Tunnel carries the highest part of the canal through the Cotswold Hills.

At 3½km (just over 2 miles) in length, it was at the time of its building the longest tunnel in Britain and it still remains the third longest ever built in the UK. King George III, having more time on his hands since losing the colonies, was suitably impressed when he visited the site in 1788. Since that time, the stretch of canal outside of the Coates Portal has been known as King's Reach. In May 2002, when I was project manager for the restoration of this waterway, we hosted a visit of HRH The Prince of Wales to the King's Reach. Prince Charles, visiting in his capacity as Patron of the Waterways Trust, was there to view the restoration work and to meet the volunteer navvies of the Cotswold Canals Trust. The Prince has been a great supporter of waterway restoration in Britain over many years.

At the other end of the tunnel is the restored Daneway Portal. The tunnel took five years to build and was opened in 1798, when boats were able to navigate inland between England's two great rivers – the Thames and the Severn – for the first time. The last-recorded commercial passage through the tunnel was made in 1911. At Sapperton, as with all canal tunnels that did not make provision for a towing path through the tunnel, horses were uncoupled as the boats entered the tunnel and were led over the hillside by the boatmen's wives or children while the boatmen 'legged' the boats through the tunnel. Lying on their backs, balanced on a wooden plank across the bows, they literally 'walked' the boat through the tunnel. Some canal companies employed 'tunnel boys' who would carry out this task in return for the payment of a couple of pennies.

National Waterway Museums

The nine iconic sites outlined above represent the birth of modern canals. The fascinating history of the inland waterways of Britain is

Queen Boadicea II trip boat on the Gloucester & Sharpness Ship Canal. (National Waterways Museum, Canal & River Trust)

recorded at the three National Waterways Museums. Collectively, the museums are home to the national waterways collections; the historic boat collections; and the national waterways archive. Each museum is housed alongside a working canal or dock, in an historic building and in a location rich in waterways heritage.

Beside the River Mersey and the Manchester Ship Canal at Ellesmere Port, this museum is well placed to demonstrate how the waterways helped to drive the UK's astonishing industrial revolution. The museum's fascinating displays are housed in a collection of fine Victorian buildings. Visitors can take boat trips through industrial landscapes, visit the blacksmith's shop and explore traditional narrow boats.

In rural Northamptonshire, this museum, housed in a restored corn mill, is located in the heart of the canal village of Stoke Bruerne, through which passes the busy Grand Union Canal. Inside the building will be found a treasure trove of stories, exhibits and collections that explain and bring to life the rich world of Britain's canals. Visiting children are encouraged to put on traditional costume and are asked to try to stand still for long enough so that a

timeless image might be captured. Outside, the entire village is alive with canal heritage. So, whether you're after a pit stop for coffee or a memorable day out, Stoke Bruerne is a delightful place to visit.

In the heart of Gloucester's historic docks, the waterways museum is housed in a handsome Grade II listed warehouse. As at its two sister locations, the museum recalls the fascinating story of the thousands of kilometres of Britain's waterways and, in addition, helps to capture the life of a working commercial dock. There are hands-on displays and archive films to watch, and on the museum quaysides there are many interesting exhibits, often demonstrated by the volunteers who restore and maintain these artefacts. Across the dock from the museum site is the ship repair yard of T. Nielsen & Company, which specialise in work on traditional wooden sailing vessels. In 2014, the yard was awarded a contract to work on Admiral Lord Nelson's Portsmouth-based flagship, HMS *Victory*. I have never visited Gloucester when there hasn't been an old pilot cutter or large square-rigged sailing ship in dock for repair. Within the museum, one can see and climb aboard a range of different craft, including an impressive steam dredger (when in steam, a visit below decks to see the mighty steam engine at work should not be missed).

To complete another perfect day out, a cruise can then be taken along the Gloucester and Sharpness Ship Canal on either of the two passenger boats operated by the museum. My particular favourite is *Queen Boadicea II*, known affectionately as *QB2*; she is of particular historic importance as she is one of those brave Dunkirk 'little ships' that assisted with the evacuation of troops from the beaches of Dunkirk in May 1940.

This short visit to these famous canal structures and to the three National Waterways Museums provides an insight into canal engineering, which, in most cases, occurred more than 100 years before the construction of the Panama and Suez Canals.

9

FORTRESS GIBRALTAR

Gibraltar has been a place of conflict for much of its historic past and many of those conflicts have influenced the development of Fortress Gibraltar, a defensive citadel I know well, having served there for three years as chief civil engineer to the Ministry of Defence.

Gibraltar's 'recorded' story begins in AD 711 when Tāriq ibn Ziyād, the Berber leader of the Moorish army, sailed from North Africa and landed on the sandy isthmus connecting the 396m-high mountain with the southern shore of Spain. Within a few months, the Muslim invasion of Christian Spain was completed and, in order to secure his line of communication across the 24km-wide strait, Tāriq's engineers constructed a castle on the north-western slope of the massive rock overlooking the isthmus. The rock was named Jabal al Tāriq (Rock of Tāriq), from which its English name, Gibraltar, is derived. It is extremely doubtful that Gibraltar was anything more than a defence outpost until about 1160 when the Almohad monarch, Abd al-Mu'min, founded a city on the peninsula – a city of mosques and palaces with elaborate water channels, engineered to convey natural water from the upper rock to the residencies and gardens below.

In 1309, the Spanish drove out the Moors and captured the Rock, holding it for just twenty-four years until it was recaptured by the Moroccan monarch Abdul 'Hassan.

Most, if not all, of the extant Moorish remains in Gibraltar stem from that period. Superbly engineered, these remains are part of the current tourist trail and include the Moorish Castle, Moorish Baths and various defensive works. Gibraltar was heavily refortified as a 'Citadel of Islam' but, despite vast expenditure, the fortifications proved insufficient to prevent the Rock's final fall from Muslim occupation in August 1462 when, on the feast of St Bernard, Spain again took Gibraltar. St Bernard subsequently became Patron Saint of Gibraltar.

The Spaniards held the Rock from 1462 until 1704. Three years before 1704, England and Holland had joined with Austria, the

Fortress Gibraltar viewed from an approaching ship.

Holy Roman Empire and Portugal in an alliance for a war against France and Spain. The War of the Spanish Succession started when it was realised that Charles II of Spain had no obvious heir to his throne. Louis XIV of France and Leopold I of Austria both claimed equal legitimacy in their right to Spanish territory. All of the major European empires tried to tip the balance of power their way, regularly switching sides. The war dragged on until the treaties of Utrecht and Baden finally put Philip V on the Spanish throne. During this war, an Anglo-Dutch fleet commanded by Vice Admiral Sir George Rooke arrived in the Bay of Gibraltar to set ashore a landing force of about 1,800 British and Dutch marines with the intention of capturing the Rock from Spain. Within three days, the defenders found that opposition was hopeless and on 24 July 1704 they surrendered.

Gibraltar was formally ceded to the British Crown under the Treaty of Utrecht in 1713, the essential words of the treaty being:

The Catholic King does hereby, for himself, his heirs and successors, yield to the Crown of Great Britain the full and entire propriety of the town and castle of Gibraltar, together with the port, fortifications, and forts thereunto belonging; and he gives up the said propriety to be held and enjoyed absolutely with all manner of right for ever, without any exception or impediment whatsoever.

Despite this unambiguous wording, Spain did not give up hope of recapturing the Rock.

Since that time, the civilian populations of Gibraltar and the British forces stationed there have been subjected to occasional sieges. The Great Siege, as it later became known, started on 13 September 1779 when the first gun was fired in the long struggle against a large Franco-Spanish army commanded by the Duc de Crillon. The British governor at the time was General Augustus Eliott, under whose tireless and able leadership garrison engineers excavated huge caverns and tunnels from which to defend the fortress. Although outnumbered by four to one, the garrison held out for three years, seven months and twelve days.

During the Battle of Trafalgar of 1805, which was fought in the sea area outside of the Spanish port of Cadiz, Admiral Lord Nelson was killed when a sniper's bullet fractured his spine. His body was

Gibraltar Post Office special issue stamp commemorating the 175th anniversary of the death of Vice Admiral Horatio Nelson. (Gibraltar Post Office)

Gibraltar Post Office First Day Cover celebrating the 150th anniversary of the Gibraltar police force. (Gibraltar Post Office)

brought back to Gibraltar aboard the battle-scarred HMS *Victory*. The Gibraltar Post Office commemorated this historic event 175 years later; its special-issue postage stamp depicted *Victory* being towed into Gibraltar by HMS *Neptune*. *Victory* anchored in Rosia Bay – one of the earliest examples of naval harbour construction in Gibraltar – and Nelson's body was taken ashore to be preserved in a barrel of rum for the journey back to England. According to naval folklore, when the barrel was opened in England it was considerably less than full of rum, giving rise to the story that sailors aboard *Victory* had been unwilling to let a little thing like a decomposing dead admiral stand between them and their daily tipple – a story that led to the naval expression 'tapping the admiral' or getting an unauthorised drink of rum via a surreptitious straw.

Although there have been no further military attempts to capture the Rock from the British since the fortress gates were opened at the end of the Great Siege on 12 March 1783, there have been further sieges. Complete closures of the frontier have been imposed by Spain in order to try to force the people of Gibraltar to renounce British sovereignty. The most recent of these sieges, which lasted for sixteen years, was between 1969 and 1985, a period that included my three-year tour of duty. Since 1985 there have been many occa-

sions when the Spanish authorities have imposed restrictions at the border, designed to cause frustration to the people of Gibraltar wishing to cross the frontier. Gibraltarians are fiercely British; it is not that they dislike Spain or the Spanish people, but they like their current situation and want things to remain as they are. This steadfastness has been demonstrated in referenda. There is serious concern in Gibraltar that Brexit might be used as an excuse to impose further border restrictions.

Incidentally, the Royal Gibraltar Police Force celebrated its 150th anniversary in 1980. Like so many other local historic events, the occasion was marked by an issue of new stamps from the Gibraltar Post Office and a First Day Cover. I collected all of the First Day Covers issued between 1978 and 1981, not least because the designer of the stamps and many of the commemorative covers was one of the Gibraltarian staff employed in my civil engineering drawing office; Mr A.G. 'Freddie' Ryman was a very talented draughtsman indeed.

Adolf Hitler's directive No. 18 of 12 November 1940 detailed the four stages of Operation Felix – his plan to capture Gibraltar. German troops were to mass in southern France and cross through Spain to attack across the isthmus. Hitler was, however, unable to reach agreement with the Spanish dictator, General Franco, who

Restrictive air space at RAF Gibraltar. (Author; Google Earth)

insisted that Spanish troops should lead the attack. Franco also demanded that German troops should bolster defences in the Canary Islands, which he was convinced the British would try to take as a reprisal for the loss of Gibraltar. British Gibraltar was, seemingly, saved by the intransigence of the Spanish dictator. Contingency plans had, nevertheless, been drawn up by British forces as part of the top-secret venture Operation Tracer. In the event of an imminent capture of Gibraltar by enemy forces, a number of military personnel were to be sealed into newly engineered chambers inside the Rock. The occupants of these chambers would be provisioned for at least a year and their task would have been to transmit by radio the movements of enemy shipping passing through the Gibraltar Strait. The plan was so secret that only rumour persisted long after the war. It was not until the discovery in 1997 of previously unknown chambers beneath Lord Airey's Battery, together with research at the Public Records Office, that Operation Tracer was finally confirmed. During his visit to Gibraltar in 2008, 93-year-old Dr Bruce Cooper, Surgeon Lieutenant (Retired) Royal Navy, the last surviving member of the 'Stay Behind Squad', was able to confirm that the chambers discovered in 1997 were in fact the secret chambers into which he and his colleagues were to have been sealed.

In both world wars, the Rock played a strategic role in the anti-submarine campaign of the Gibraltar Strait. My tour of duty in Gibraltar was within the Cold War years and new projects generated at that time played their part in keeping that particular war 'cold'. Due to the Official Secrets Act, it is not possible, even now, to provide any detail of those projects. Suffice to say that in both world wars the Rock was a strategic location for monitoring the position of enemy submarines – and some things just do not change!

The Second World War military planners, concerned that Hitler would deploy submarines to block the Gibraltar Strait, concluded that a fully operational airfield was needed at Gibraltar, but the small grass air strip completed in March 1936 at a cost of £573 was considered not fit for purpose. A decision was taken to build an airfield on the sandy isthmus below the north face of the Rock. Thousands of tons of crushed stone would be required for its construction, some of which came from newly opened quarries and some of which was removed from inside the Rock where mining operations were under way, extending the tunnel system. General Eisenhower established his headquarters within these tunnels from where he commanded Operation Torch, the combined British and American invasion of North Africa of November 1942.

The many kilometres of tunnel inside the Rock also fell to my responsibility. Some had been opened as public roads, many were still used by the military and some contained reservoirs to store precious drinking water. These reservoirs were constructed at the time of an earlier siege and were used in association with the unique 'water catchments' on the east side of the Rock.

The airfield at Gibraltar was then (as now) operated by the RAF. The last of three extensions of the solid-surface runway, which was begun in late 1939, was completed in 1955. With a final length of 1.8km, the runway is able to handle most current RAF aircraft. Defence cuts over many years have dramatically reduced the size of RAF Gibraltar. Commanded now by a wing commander, the airfield is run by fewer than forty-five service men and women. Although a small unit, RAF Gibraltar continues to play a crucial role within British Forces Gibraltar, being maintained as a forward operating base for UK aircraft. It was extensively used in this role during the Falklands conflict and in the Gulf War. Its position at the western end of the Mediterranean also makes it an ideal supporting base for major NATO exercises in the Mediterranean and Iberian peninsula areas.

As well as being a military airfield it is also an international airport, serving scheduled civilian flights to and from the UK, Spain and North Africa. With prohibited Spanish airspace immediately to the north of the airfield and the 415m-high rock only ½km to the south, landing there can pose a unique challenge when particular wind speeds and direction causes severe turbulence around the top of the Rock. The main runway, projecting outwards into the sea at its western end, has sea at both ends and is unusual in having a four-lane public road (the main road into Spain) crossing the middle of it. Because it is relatively short, the runway surface has had grooves mechanically cut into it, in order to improve aircraft stopping distance and to reduce the risk of aquaplaning. Being aware that these 10mm x 10mm grooves, spaced at 50mm centres, might induce more significant cracking, my regular inspections of all airfield pavements included a close scrutiny of the grooving. For some years now, the Gibraltar government has been planning to remove the road crossing by building a new road tunnel at the extreme eastern end of the runway. However, following completion of the preliminary works (the removal of 27,000 tonnes of contaminated sand) all activities have, for some reason, been suspended.

My technical teams carried out daily inspections of the runway, taxiways and parking aprons. The aprons were used by visiting aircraft and by the Gibraltar Airways Vickers Viscount passenger plane (extreme left of the photograph above). The Viscount ran a regular and essential air ferry between Gibraltar and Tangier, with occasional weekend excursions to other destinations in Morocco

RAF Gibraltar aircraft parking apron.

and Portugal. Remember that, with a closed frontier, there were very limited recreational routes away from Gibraltar. Seeing *Yogi*, as the aircraft was affectionately known, standing on this particular part of the apron reminds me of a call I received from the senior technical officer at the airfield explaining that he had filled in a shallow depression in the taxiway next to *Yogi*'s stand three times, and that within a day or so of each filling the surface had once again sunk, recreating the depression. My recommendation was that we needed to excavate carefully through his three fillings in order to determine just what, beneath the taxiway, was causing the problem. The clue to the cause of the problem lay in the fact that prior to the Second World War the area currently occupied by the airfield was used as a racecourse. Because this was the only large area of flat ground in the colony, when a prized racehorse died, the horse was buried on the racecourse. What we discovered at the bottom of our depression was a horse! It had long since been dead, of course, but the ribcage of the animal had, for decades, been supporting the tracking nose wheels of aircraft until its arched bones finally gave way.

In June 1940, Italy's dictator Benito Mussolini joined forces with Hitler and declared war on Britain and France. Two months after that declaration of war, an Italian campaign against Gibraltar began. Two 620-ton Italian Mediterranean-type submarines – *Scire* and her sister ship *Gondar* – underwent special modifications at La Spezia Naval Base on the north-west coast of Italy. Each submarine had its gun removed and conning tower reduced in height. Three steel cylinders were fitted to the upper deck of the submarines, each designed to carry a single 'human torpedo'. These torpedoes were nicknamed by the Italians as '*maiali*' (pigs) because they looked like a swimming pig and were just about as easy to manoeuvre. The submarines transported the torpedoes and charioteers to a point as close to their final destination as possible and the torpedoes were then released underwater. Two charioteers then sat astride each torpedo in order to pilot it to its target. Each torpedo had a detachable 300kg explosive warhead.

The first operation for the newly converted submarines was a dual attack on the two British naval bases at opposite ends of the Mediterranean: Alexandria and Gibraltar. *Gondar* sailed east and *Scire* left La Spezia on 24 September 1940 to sail west to Gibraltar. When *Scire* was only 80km out from Gibraltar, Italian Supreme Naval Command in Rome cancelled the operation, as intelligence sources reported that there were no Allied ships in Gibraltar Harbour.

The following month, a second operation was launched. *Scire* left La Spezia to come to rest on the bottom of Gibraltar Bay near the mouth of the Guadarranque river at 1.30 a.m. on 30 October 1940. Italian Naval Command had signalled that two British battleships were alongside in the harbour, one of which was HMS *Barham*. At 2 a.m., the three crews sat astride their torpedoes and set out to attack the battleships, neither of which was actually damaged. The explosion of a warhead within the harbour the following morning

alerted the British to the near success of the Italian operation. The means of delivery of the attack became all too apparent a little later when a wandering torpedo released from a Spanish beach by its crew the evening before beached itself near the Rock. From then on explosive charges were regularly dropped into the harbour and into the waters of the anchorage outside of the harbour.

In March 1941, the Italian Special Underwater Weapon Unit adopted a new procedure to get torpedo crews to their Gibraltar targets more rested and, therefore, more capable of achieving a successful attack. Instead of travelling for days in the cramped conditions of the parent submarine, the crews were to be flown from Italy to Cadiz, where they would rendezvous with *Scire*. The Italian tanker *Fulgor*, which had secretly been equipped with stores, was conveniently moored in Cadiz Harbour. On the evening of 24 May 1941, *Scire*, carrying three torpedoes, slipped silently alongside *Fulgor* to embark the now rested torpedo crews. The submarine departed before daybreak on the short sea passage to Gibraltar. Italian intelligence reported that there were no British warships in the harbour and so they instructed that merchant ships in the anchorage were to be attacked instead. Once again, events did not go entirely the way the attackers would have wished and no Allied ships were damaged. Although the actual operation had been another failure, Italian High Command were advised that it had been a useful training exercise, demonstrating that the new crew delivery system via *Fulgor* in Cadiz and their departure by air from Seville after the deployment had been a great success. A series of successful raids using this route followed.

In 1942 another change of operation took place. The 5,000-ton Italian tanker *Olterra*, which had been scuttled off Gibraltar in 1940 to prevent her from falling into British hands, was refloated. She was then moved to a berth at the end of the outer pier of Algeciras Harbour just across the bay from Gibraltar. Italian Naval officers and engineers, disguised as tanker engineers, began the work of converting *Olterra* into a new human torpedo base. A hinged door was cut through the side of the hull to allow manned torpedoes to exit from a secret underwater compartment created by flooding the forward part of the ship. The torpedoes and other specialist equipment were smuggled past unsuspecting Spanish customs officers disguised as spare parts (such as piping and machinery) for the 'legitimate repair' of the ship. So, right under the noses of the Royal Navy in Gibraltar, a secret Italian naval base was being set up to carry out operations against Allied ships using manned torpedoes that were covertly deployed from below the waterline of the innocent-looking *Olterra*, moored just a few kilometres from Gibraltar.

Ready for their first Olterra Operation on 6 December 1942, the Italians could not believe their luck, for, on the evening before, British Naval Squadron Force H arrived in port. The Italians planned to use three torpedoes: the first would target the battleship *Nelson*; the second would target the aircraft carrier *Formidable*; and the third would target the aircraft carrier *Furious*. That evening, as the crews set out from *Olterra* via the underwater hinged door, British searchlights high on the Rock played constantly back and forth across the bay and explosive charges were dropped into the water every three minutes. Submarine nets barred both entrances to the harbour, which were also covered by guns. The attack was a disaster for the charioteers; three were killed and two were arrested, with only one returning safely to *Olterra*. Undeterred by the loss of five men, more crews and torpedoes were delivered to the ship. A decision was, however, made by Italian High Command that, as Gibraltar Harbour defences were judged too difficult to penetrate, all future targets would be limited to merchant ships lying at anchor in the bay. Further successful missions followed until, on 8 September 1943, Italy signed an armistice with the Allies ending the Italian campaign to sink Allied shipping at Gibraltar. Enemies they might have been but, on reflection, they could also have been considered brave warriors who caused considerable disruption to Britain's war effort.

German Navy Deutschland-class heavy cruiser *Admiral Scheer* at Gibraltar naval base headquarters. (Official US Navy photo NH 59.09 9664, via Wikimedia Commons)

Throughout both world wars, the naval dockyard in Gibraltar worked at full pressure repairing British and Allied warships. In between these conflicts major units of the German Navy paid courtesy visits to Gibraltar. *Admiral Scheer*, a Deutschland-class heavy cruiser (sometimes termed 'pocket battleship'), innocently moored alongside the naval base headquarters in 1936; no doubt she departed with much military intelligence gathered during her courtesy visit. The ship was subsequently sunk during an Allied bombing raid on Kiel in 1945.

This brings us neatly to Royal Naval infrastructure in Gibraltar. In 1894, the Admiralty began the preliminary operations of a comprehensive project for the enclosure and defence of the harbour and the extension of the dockyard to include three large dry docks. These works and other improvements continued until 1905.

Quarries to win stone for the rubble hearting of the quay walls and for making concrete were opened up in Gibraltar at Catalan Bay and at North Front, where a large concrete-block-making yard was established. Dressed stone came chiefly from the Europa Quarry and sand was obtained from the slopes on the east side

HMS *Londonderry* (F108), a Rothsay-class Type 12 frigate in dry dock in Gibraltar.

HMY *Britannia* at Gibraltar naval base headquarters, August 1981.

of the Rock. During the course of the project, 207,000 tons of cement were imported into Gibraltar. Granite for sets and quay-wall edges came from Cornwall and limestone ashlar for quay walls, docks, dry docks and buildings mainly came ready-dressed from Spain. The Spanish stone was delivered to Algeciras by railway and transported in lighters across the bay to Gibraltar. The landing pier and quarries were linked to the harbour works by a metre gauge railway system built around and through the Rock. This railway disappeared soon after completion of the dockyard, with no evidence of its tracks to be seen anywhere in Gibraltar, but, whilst undertaking some road works in 1980, we uncovered a buried section of the track. The rail and sleeper arrangement was unusual in that the rail was supported on short cut lengths of interlocking steel pile sections which, in turn, were supported on square pads of concrete. All of the dockyard infrastructure, which was completed at the turn of the twentieth century, is still in use today, more than 100 years later.

During the time of my tour, the dry docks were used for complete refits and repair of HM warships. HMS *Londonderry* (F108),

a Rothsay class Type 12 frigate, underwent such a refit. She was built in 1960 by J.S. White of Cowes on the Isle of Wight – sadly a lost name in the annals of British ship building. Major units of the fleet also visited Gibraltar on a regular basis and this included HMS *Invincible*, which called in on her maiden operational deployment from the UK. She was only in harbour for a couple of days but attracted a great deal of attention. The Combined Engineering Society in Gibraltar, of which, as chief civil engineer, I was a member, tried to arrange an official visit to the ship but this could not be fitted into her very tight programme. I have to report, however, that the Civilian Wives Club was afforded that privilege and I still cannot understand naval priorities, preferring a group of young ladies to a bunch of engineers! Another major unit visiting in June 1980 was the aircraft carrier HMS *Bulwark*; she underwent work in No. 1 Dry Dock before 'paying off' into Indian Navy command.

The Gibraltar squadron is now the only resident seagoing Royal Naval unit in Gibraltar. It is an operational frontline squadron currently consisting of two 16m patrol launches and three 6.5m rigid inflatable boats. Based in a purpose-built headquarters,

Camber deck at Gibraltar's Royal Navy dockyard.

Severe corrosion to the underside of the camber deck above dry-dock sliding caisson.

the nineteen-person team is operational throughout the year in order to maintain the security and integrity of British Gibraltar Territorial Waters.

Royal dockyard tasks in 1982 included work to *Uganda*. Between 1952 and 1967 *Uganda*, a liner of the British–India Steam Navigation Company, operated between London and East Africa with regular calls into Gibraltar. In 1968 she was converted into a very popular educational cruise ship for schools. It was while undertaking a school's cruise that she was requisitioned by the Ministry of Defence and sent to Gibraltar for conversion into a hospital ship for service in the Falklands War.

The Gibraltar passenger-cruise terminal is located on the North Mole, external to the naval dockyard. In my time, it provided berths for all visiting passenger ships, except for P&O's then flagship *Canberra*. Due to her draft, which was greater than other ships of the day, she was allocated a deeper berth on the South Mole within the naval base. It was part of my team's responsibility to ensure that the berth was kept dredged and in good order for that important and fairly frequent visitor.

Perhaps the most prestigious occasion at the Gibraltar naval base that I recall was the arrival in July 1981 of the *Royal Yacht Britannia* to embark a certain young couple for their honeymoon cruise. There was a lot of pleasure-boat activity within the harbour on that day. The Prince and Princess of Wales had travelled from the UK to land at RAF Gibraltar in an aircraft of the Queen's Flight, which was piloted for most of the journey by Prince Charles. Due to the large number of guests that the Port Admiral had intended to invite to watch the royal departure from the front balconies of the naval base headquarters (the building with the tower on the left of the harbour scene, opposite), I had the invidious duty to advise him that numbers would have to be severely limited due to the then serious corrosion of the beams supporting the balconies. Somewhat surprisingly in the circumstances, my wife and I were still included on his shortened guest list, perhaps because he wanted his civil engineer on hand if things did begin to creak and bend! In the event, all was well, and the day went off without incident; at least on the Rock it did, but out at sea, King Juan Carlos of Spain cancelled a planned visit to the

East Side Ammunition Jetty, Gibraltar, under attack from an easterly storm.

royal yacht due to the Spanish government's official objection to the 'high-profile use of Gibraltar'.

Corrosion of structural steelwork in the salt-laden airs of Gibraltar was a major issue with which I had to deal, and in the dockyard it was not just confined to the prestigious naval base headquarters tower. Each of the three huge sliding caisson dry-dock gates were housed under a camber deck. Hinged at one end, the deck is lifted by rams acting on a steel-box section protruding from each of its sides. Once raised, the caisson beneath is then able to slide out to close off the dry dock. The structure, essentially a beam and cantilever plate girder, looks sound enough when viewed from above, but underneath it was quite a different story. Load-bearing stiffening webs had all but disappeared, eaten away by corrosion. The British Standards Institution design codes demanded that this type of stiffener should be 'fitted' between the top and bottom flanges of the girder. Clearly, they were no longer fitted and, for much of its length, there was no bottom flange plate left to fit them to. Needless to say, following my first inspection, I condemned the structure. Perhaps this area was not easily accessible for routine maintenance painting, but clearly more effort should have been made over many previous years to give this essential structure some protection from the corrosive salt-laden air.

Meanwhile back in the harbour, the structure that caused me my biggest 'political' headache, due to its corroded condition, was Viaduct Bridge. As can be seen on the map on page 145, which indicates the layout of the waterfront as it was in the 1980s, Viaduct Bridge provided the only link between the commercial docks and the city, a link that was critically essential at a time when the Spanish frontier was closed, the docks being the only point of entry for bulk cargoes into Gibraltar.

Viaduct Bridge was a classic structural design for the early 1900s, with the principal beams being prefabricated from flat steel plates and angle sections riveted together to form the girders.

The total length of the bridge was 110m, being divided into six spans. It was carried on seven pairs of 10m-long concrete-filled cast-iron cylinders, half the length of which was sunk into the seabed. On my first inspection of the underside of the bridge, I noted with some horror that the condition of the steel-support girders was very similar to those under the dockyard camber deck. Given its condition, quite how the bridge was managing to support freight-container traffic loads imposed upon it was a mystery to me. It was, therefore, with some trepidation that I approached the Governor of Gibraltar, General Sir William Jackson, to explain how Viaduct Bridge was in imminent danger of collapse. If it did collapse, we both, of course, realised that Gibraltar would be cut off from the rest of the world as far as freight was concerned. Emergency plans were immediately put into place to try to reduce the risk of collapse and minimise the risk to users. Only one vehicle, of a maximum of 10 tons all-up weight, was permitted on the bridge at any one time and vehicles were to move no faster than at walking pace down the centre line of the structure. This was to ensure that loads were spread evenly between the two fragile

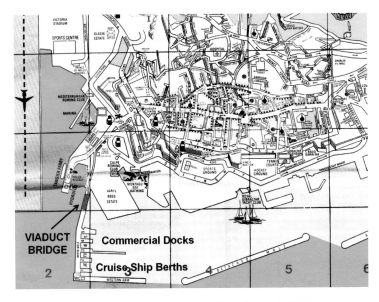

Map showing Gibraltar Commercial Docks in the early 1980s.

girders. A twenty-four-hour police post was established at each end of the bridge to ensure that these rules were obeyed. I was able to borrow two-dozen dockyard welders to begin a process of temporary repair. By this means, we were able to extend the life of this essential crossing until a permanent rock causeway could be built alongside. The harbour area around the bridge has since been in-filled and, whilst the only route from the commercial docks and passenger cruise terminal to the city passes over the location of Viaduct Bridge, nothing is now visible.

Elements of the wider building stock in Gibraltar also exhibited problems with corrosion. Although the majority of buildings on the navy and RAF estates were fairly modern operational buildings, those on the army estate tended to be older, many of them of historic value. Typical were Victorian barrack accommodation blocks; indeed, my own office, New Mole House, was once a cavalry barracks (today it houses Gibraltar police headquarters). Corrosion was sometimes a serious problem in many of these older buildings. Whilst carrying out an inspection at the 'Old Naval Hospital', then in use as officers' married quarters, I noticed that corrosion at the

foot of a steel column had been disguised by the application of a coat of paint. On scraping away the paint with a penknife, most of the steel across the entire section of the column (which supported a roof structure above) was also easily removed. Another example of poor maintenance practices over many years.

An incident that took place at Gibraltar naval base just over a quarter of a century before my tour had a direct effect on two of the projects with which I was involved. In 1951, the naval armament vessel RFA *Bedenham* arrived in Gibraltar carrying 500 tons of mixed ammunition. A barge moored alongside *Bedenham* to assist with unloading caught fire. This, in turn, caused *Bedenham* to catch fire, leading to a devastating explosion, earning it an infamous place in the history of Gibraltar. The explosion sent up a mushroom cloud of smoke that could be seen from North Africa across the Strait of Gibraltar. Pieces of the ship were actually blown right over the top of the Rock, leaving debris covering much of Gibraltar. Nearby vessels were sunk and adjacent dockyard buildings destroyed. The cathedral in the centre of town was severely damaged, as was the Governor's Residence in Main Street; in fact, there was hardly a building in Gibraltar that did not suffer some damage. The explosion caused seven deaths and hundreds of injuries. More destruction was caused by this one incident than in all of the air raids of the Second World War. As a direct result of this devastation, a decision was made by the Admiralty that ammunition ships would never again be unloaded in the harbour so close to the city centre; instead, a jetty was to be located on the remote east side of the Rock for all future handling of ammunition ships.

East Side Ammunition Jetty, sited directly below a sheer 365m cliff face, is accessed via two road tunnels: the long and straight 'Dudley Ward Tunnel', which carries the main Sir Herbert Miles Road around the Rock; and, on a falling gradient, actually passing under the main road tunnel, the appropriately named 'Dog-Leg Tunnel'. This tunnel makes a T-junction inside the Rock with

Dudley Ward Tunnel and exits directly onto the jetty. It might come as a surprise to those who only know the Mediterranean in the kinder summer months that in the winter, easterly storms can be quite ferocious. With the jetty located on the exposed eastern coast of Gibraltar, and with a clear fetch of some 1,600km of open sea all the way from Sicily, huge waves would build up and, consequently, the jetty regularly suffered severe storm damage. When an easterly gale had been running for some days, waves would frequently run along the jetty deck and drive up for some distance inside Dog-Leg Tunnel – not, of course, a good time to be unloading explosives. The worst of the damage was caused at the root of the jetty, where the unbelievable explosive force of the waves, when trapped between the jetty and the Rock, literally blew the jetty apart.

The Ministry of Defence civil engineering department had already gone out to tender at the time of my arrival with a project to rebuild the shore end of the jetty with the same design specification that had been used both for the original construction and for subsequent repairs. The tender process was suspended so that I might reconsider the design. Since the sea had consistently demonstrated its intention to remove this part of the jetty, my redesign of the project was to do exactly that! A series of heavy reinforced concrete beams had first to be manufactured. A casting yard was established on a ledge high up on the Rock, close to the southern entrance of Dudley Ward Tunnel (local topography did not provide anywhere closer to the jetty). After being cast by a civilian contractor, army manpower, plus some kit borrowed from the RAF, enabled these heavy concrete beams to be manoeuvred onto the public road and into Dudley Ward Tunnel, to later make that very tight turn into Dog-Leg Tunnel inside the Rock and so on down to the jetty. It was the dimension of Dog-Leg Tunnel, with its tight bends and right-angled junction between the two tunnels, that had set the criteria for the maximum length of beams we were able to consider. Once safely out on the jetty, the beams were

Gibraltar water catchments under repair in the late 1970s.

offloaded. The military kit was then released back to its everyday use at the airfield, the soldiers returned to their normal duties and the civilian contractor once again took over construction. My revised design called for the new concrete beams to be fixed into place, with one end of the Rock anchored just inside the tunnel mouth and the other end fixed to what was to become the new shore end of the jetty. Once the beams were securely fixed, what remained of the damaged root of the jetty beneath was removed. Destructive wave energy was no longer trapped between jetty and shore but was allowed to be safely dissipated between them and up through the gaps between the concrete beams, with no resulting damage to the structure. When required for unloading ammunition (which, of course, only takes place in calmer weather) specially fabricated steel plates, which were stored inside the tunnel, were laid between the concrete beams to complete the deck surface.

I mentioned earlier that the *Bedenham* explosion all those years ago had a direct effect on two of the activities with which I was involved. The second affected activity was the routine mainte-nance dredging of the harbour. Because of the possible presence of unstable, unexploded ordnance on the seabed in the area where *Bedenham* met her watery end, dredging contracts had to be very carefully designed, planned and monitored.

Closed frontier situations with Spain have made it necessary for Gibraltar to be self-sufficient in the provision of its public water supply. As there are no subterranean sources available, all potable water had to be collected from rainfall or produced on the Rock. When first introduced to the Glen Rocky Distillery, with its distinc-tively Scottish name, I must admit to being just a little disappointed to learn that we manufactured drinking water and not whisky at the facility. Glen Rocky was one of three distillation plants operated at that time by the MOD, supplying drinking water to the military estate and to all visiting ships. The average annual consumption was around 600,000 tons, of which approximately 10 per cent was from rainfall.

The 'water catchments' constructed in 1908–09 were once an easily recognisable feature on the east side of Gibraltar and were clearly visible from the decks of passing ships. Thirteen hectares of a steeply inclined natural, windblown sand slope had been sheeted with corrugated iron in what was, in construction terms, essentially a vast roof. The sheets were supported on timber rafters, which, in turn, were supported by timber piles driven into the sand slope. The corrugated-iron sheets were coated with a cement wash to improve corrosion protection, giving the false impression when viewed from a distance and in bright sunshine that this was a concrete struc-ture. Two water channels running horizontally across, one halfway down the slope and the other at the base of the slope, channelled water into reservoirs excavated within the Rock. The 'catchments' swallowed up a sizeable part of my annual maintenance budget. In the photograph opposite, maintenance work was being under-taken 225m above sea level, such work being necessary to replace life-expired parts of the timber structure or to repair holes in the surface caused by falling rocks. Holes had to be repaired without delay, as they concentrated water flow onto the sands below, causing washout leading to the undermining of the structure. Some years after my tour ended and, following the commissioning of further desalination facilities, a decision was made to remove all of the man-made structure from the sand slope. Consulting engineer Gifford & Partners developed a scheme whereby removal of the structure and stabilisation of the sand slope were carried out simultaneously. A three-and-half-year contract, involving the installation of some 11,000 soil nails combined with the use of seeded coir geo-textiles, was completed in 2003. The sand slope has turned progressively more and more green each year, changing forever the appearance of the east side of the Rock as viewed from passing ships, which had, for the previous 100 years, appeared snowy white in the early morning sunshine.

Gibraltar was a most interesting posting for a civil engineer because of the amazing variety of work within a fairly confined geographic area. From building structures to airfield pavements, rock stability and tunneling, harbour works, water supply, highway maintenance and masts and towers – all made more challenging by a closed land frontier. Restricted access placed additional demands on the logistics of material supply and the management of a dis-parate labour resource, many of the operatives commuting on a weekly basis from Morocco. So lots of problems to solve!

Cutter-suction dredger *Castor*, which was used to excavate the tunnel trench for the Øresund crossing. (Van Oord)

10

CROSSING AND TAMING THE BALTIC SEA

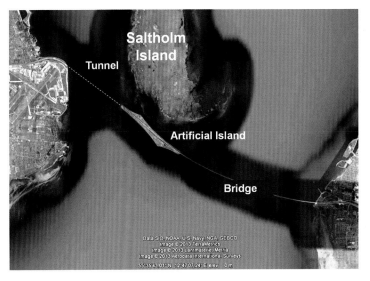

Google Earth image of the Øresund Link between Copenhagen and Malmö. (Author; Google Earth)

Two major twenty-first-century civil engineering structures have changed for all time the way that ships navigate the Baltic seaways. The first of these is located at the very entrance to the Baltic, known as the Øresund Road and Rail Crossing, and provides a fixed link between Denmark and Sweden. The second is the St Petersburg storm barrier built across the head of the Baltic in the Gulf of Finland to protect St Petersburg from severe storm surges and, through which, ships have to navigate in order to gain access to the city.

The Øresund Road and Rail Crossing is a hybrid part-tunnel and part-bridge link joining Denmark's capital, Copenhagen, to Malmö in Sweden. The Skanska-led consortium, Sundlink Contractors, built the $1.3 billion(US) crossing for the Øresund Bridge Consortium, which owns and operates the bridge. Constructed between 1995 and 2000, this 'design and build' contract was completed six months ahead of schedule.

Crossing two major shipping channels, the Flinte Channel is catered for by a high-level bridge and navigation in the Drogden Channel is unaffected as the crossing is contained in an immersed concrete tunnel. In order to facilitate transition from bridge to tunnel an artificial island was created linking the two elements.

The island provides sufficient space for both road and rail to descend from the high-level viaduct down through the body of the island to disappear into the tunnel beneath its surface. On the foreshore, adjacent to Copenhagen International Airport, an artificial peninsula was grafted onto the original shoreline in order to provide space for the tunnel to emerge from the depths without interfering with operations at the airport, which had been built right up to the original shoreline.

So why did the design engineers choose a hybrid part-tunnel, part-bridge solution, particularly as a detailed cost analysis had demonstrated beyond all doubt that a bridge would have been the most cost-effective way of linking Denmark to Sweden? Consideration had to be given to operations at Copenhagen Airport. A bridge across the Drogden Channel would have had to have been a high-level structure to enable large ships sufficient headroom to track beneath. It was determined that such a structure at this location would present an unacceptable hazard to aircraft using the airport and so could not be allowed. So that was the reason why a hybrid solution was deemed necessary, but why go to the trouble of building an artificial island to link the bridge and the tunnel, when Saltholm Island lay directly on a straight line between Copenhagen and Malmö? Would it not have been a neater and more cost-effective solution to build the bridge section from Malmö to Saltholm Island and the tunnel from there to Copenhagen, thus avoiding the cost of the artificial island? That solution was ruled out on environmental grounds; Saltholm Island supports an important bird colony and is designated as a 'Natura 2000' site – one of a network of sites selected by the European Commission to ensure the long-term survival of Europe's

most valuable and threatened species and habitats. The new link was not permitted to interfere in any way with Saltholm Island.

With the form of the hybrid solution determined, let us now look in a little more detail at the main components of the Øresund crossing. The tunnel between the artificial island and the artificial peninsula is one of the world's largest 'immersed tunnels', having a portal-to-portal length of more than 4km. Most people, when they visualise a tunnel, either think of a bored tunnel like those cut into the clay beneath the streets of London for the underground railway, or they visualise the type of tunnel blasted through solid rock like those carrying roads and railways through the Alps. An immersed tunnel is constructed in an entirely different way.

The local geology at Øresund consists of glacial deposits overlaying Copenhagen limestone. First, a trench was excavated through the glacial deposits exposing the top of the limestone bed and then the excavation was carried on down into the limestone rock. All of the material removed during these excavations, together with that from the repositioning of navigation channels, was used to form the artificial island and the extended mainland peninsula. Cutter-suction dredger *Castor*, at 2,657 gross tons, is one of the largest vessels of its type in the world. Built in 1983, it was this machine that was used to excavate the tunnel trench. Equipped with a large rotating cutter head at its front end, it is designed to crunch rock at depths of up to 33m. Mounted at the end of a long arm, the head is lowered down beneath the catamaran-style double hull to the required depth of cut. Bottom material, fragmented by the cutting head, is pumped up to the surface in an abstraction pipe that is continuous through the hull of the dredger from where it is extended to the discharge point. The vessel is held in position by a 'spud leg' about which it rotates when cutting into the seabed. The vessel's two spud legs are seen in the raised position on the left-hand side of the photograph on page 148; one will be hydraulically lowered and pressed into the seabed before any cutting operation begins.

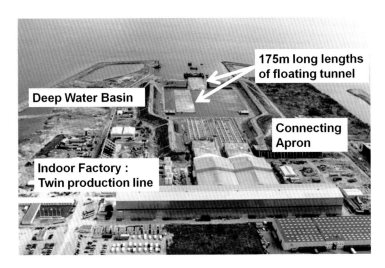

Concrete casting factory for the Øresund tunnel segments, Copenhagen North Harbour.

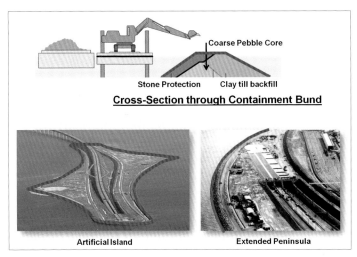

Containment bund for the artificial island and extended mainland peninsula, part of the Øresund crossing that links Denmark with Sweden.

At the same time as the tunnel trench was being cut, 12km away in Copenhagen's North Harbour, concrete tunnel segments were manufactured at a purpose-built factory. Each of two indoor production lines manufactured a tunnel segment every seven days. Newly completed segments were pushed out of the indoor factory along skidding beams onto a 'connecting apron' where eight segments were wired together. The wires were then tensioned, pulling the segments together to form a single 175m-long section of the tunnel. Whilst on the connecting apron, each end of each pair of these 175m-long lengths of tunnel was temporarily sealed off with a steel watertight wall, turning each structure into a watertight box. The connecting apron was then flooded to a depth just deep enough to allow the boxes to float and then each was moved forward, in turn, into a deep water basin where trim was adjusted to ready each length of tunnel for its 'voyage' to the tunnel site, during which time it was in the charge of four tugs. The production rate at the factory was such that two 175m-long lengths of tunnel were completed every two months.

In readiness for its arrival at the tunnel site, a precisely levelled gravel bed foundation had been added to the bottom of the exca-vated trench. Each floating length of tunnel was then accurately manoeuvred into position before being carefully lowered by allowing a controlled volume of water to enter the structure (not such a straightforward operation when you consider that each 175m-long length of tunnel was 42m wide, 8.5m high and weighed 55,000 tonnes). Each newly placed length had to be aligned with, and lightly touch, its predecessor.

Backfill material was then placed alongside the tunnel segment to fill the trench completely before a layer of rock armour was placed on top of the structure to give it protection from falling debris or dragging ship's anchors. Twenty separate lengths of tunnel were placed on the seabed in this way and, once jointing material had been pumped into place between them and the temporary steel diaphragms were removed from their ends, then seamless tubes became available for road and rail traffic.

Now let us consider in a little more detail the artificial island and the extended mainland peninsula. As I mentioned before, the seabed material excavated from the 4km-long tunnel trench, together with that removed from the relocated and deepened shipping channels

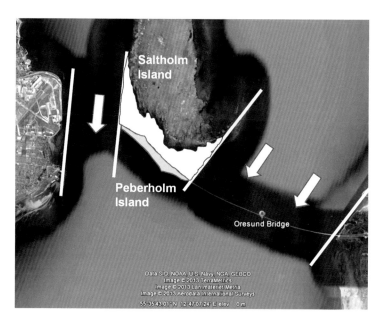

The artificial Peberholm Island in the tidal shadow of Saltholm Island, a Danish island in the Øresund Strait. (Google Earth; Author)

(all 7.5 million cubic metres of it) were used to create these new structures. However, before any of this material could be placed in position, a containment bund had first to be constructed around the perimeter of the island (which was 4km long by 0.5km wide) and along the entire new length of foreshore of the extended peninsula. Each was constructed of a coarse pebble core fronted by stone protection to prevent wave damage and was backed by clay backfill to prevent leachates from the fill material escaping into the marine environment. The bunds were built up using the dipper dredger *Chicago*, which was also used to remove material from other locations, such as the temporarily relocated shipping channels. Loaded barges were towed to the landfill sites to be offloaded by *Chicago*, which was able to handle 22 cubic metres of loose rock in one mouthful. The larger the quantity of material grabbed in one bite, the less was the percentage of silty material spilled into the marine environment.

Once the bunds were completed, the enclosed areas could then be filled with excavated material. Material was principally placed directly from the abstraction pipe on the cutter-suction dredger *Castor* via a 850mm-diameter discharge pipe supported on floats and extended right into the bunded areas of the island and the peninsula allowing the ground-up rock fragments from the tunnel excavations to be pumped directly ashore. During its construction the artificial island was formally named Peberholm.

Care for the wider environment formed a major part of the Øresund project. The Baltic Sea is the world's largest body of brackish water and it has a unique marine ecosystem, dependent on a balanced flow of saline and oxygenated water from the North Sea. The civil engineering designs aimed for a zero net impact on the marine environment by minimising changes in the flow of water through the Øresund Strait. This was achieved by constructing Peberholm Island in the shadow Saltholm Island. The project was considered a model for mitigating the environmental impacts of mega-construction projects and, for this, it was the recipient of international awards. One being the 'Outstanding Structure Award 2002 for Innovative Planning and Construction Management Techniques and Environmental Considerations' awarded by IABSE (The International Association for Bridge and Structural Engineering).

On Peberholm itself, scientists wanted to observe how the natural environment might self-generate on this barren man-made island. So, habitat creation formed no part of the project requirements. To minimise the effects of human intervention the road across the island was fenced off from the hinterland and no stopping places were provided. Shortly after completion, a variety of flora quickly became established. About 300 different plants, thirteen different bird species and a number of rare insects have now become established. Biologists were amazed to find, among the score or so species of spider found on the island, 'Eneplognatha Mordax', which actually features on Sweden's threatened spider species list. The island itself has now been added to the 'Natura 2000' listing of Saltholm Island.

The High Bridge over the Flinte Navigation Channel and the east and west approach viaducts. (Arup)

Having considered the tunnel and the landfill parts of the project, let's now consider the bridge itself. This part of the crossing has an overall length above water of just less than 8km (roughly half the total distance between Denmark and Sweden). The western approach viaduct between the artificial island and the high bridge is 3km in length. The high bridge over the Flinte Navigation Channel is just over 1km long, and the eastern approach viaduct leading from the high bridge into Malmö is just less than 4km long.

The high bridge is a cable-stayed structure with a length between towers of nearly ½km. It was designed by the global firm of consulting engineers Ove Arup & Partners Ltd, who were also the lead member of the team that won the design competition for the Link as a whole. The towers stand 204m above the water and the navigable headroom beneath the main span is 57m. Carrying a motorway and twin heavy rail tracks, it is one of the longest cable-stayed spans of its type in the world and certainly the longest in Europe. Unlike the tunnel section where the road and rail lines are located side by side, the rail tracks on the bridge are located beneath the motorway, which, itself, is carried on a concrete deck. In total some 47,000 passenger trains and nearly 9,000 freight trains uses the crossing in a typical year.

Forty-nine composite steel and concrete deck girders, supported on concrete piers slotted into concrete caisson foundations, make up the two approach viaducts. Just as the concrete tunnel components were manufactured off-site, so too were all of the pre-cast concrete viaduct supporting piers and caisson foundations, this time in Malmö's commercial docks. As each element was completed, it was transported out to site by the seagoing heavy-lift crane *Svanen*, a catamaran double-hulled vessel with a lifting capacity of 9,500 tonnes.

Perhaps surprisingly, all forty-nine of the 140m-long composite concrete and steel viaduct spans were manufactured at the Puerto Real fabrication yard of Dragados Offshore in Cádiz, Spain. Two

Seagoing heavy-lift crane *Svanen* located at Malmö Docks, Sweden. (Van Oord)

units, each weighing 5,000 tons, were placed side by side on board an ocean-going barge that departed every twenty-one days for the long journey to Malmö. At Malmö Docks, the spans were lifted from the barge and individually transported to the bridge site by the very busy heavy-lift crane *Svanen*, which then placed them upon the waiting concrete piers. Once in position, the individual bridge spans were welded together on site. Temporary cabins were erected on each pier to provide an all-weather working environment for the welders, necessary to ensure the quality of the welds – the welds being of crucial importance to the structural integrity of the viaducts, which are designed to be able to cope with the loss of one supporting pier that might accidentally be removed by ship impact.

An entirely different method of construction was adopted for the high bridge across the Flinte Channel. The concrete foundation caissons for the bridge towers were, like all of the viaduct foundation caissons, pre-cast ashore in Malmö. However, these two caissons were far too large and heavy to be lifted and transported

Øresund's high bridge towers with the artificial Peberholm Island in the background. (Arup)

to site, so they were designed as floating concrete structures. They were built in a Malmö dry dock, and when completed the dry dock was flooded, allowing each caisson to be towed by tug to the bridge site where they were sunk into position on a prepared seabed.

Because of the huge dimensions of the concrete bridge towers (at the time of build they were the highest freestanding bridge pylons in the world) they could not be pre-cast but had to be cast in situ in their final position using a slip-forming concrete technique. At the top of each of the towers, a mobile enclosure known as a climbing shutter was built. Wet concrete was then lifted from sea level by crane in a skip and poured into a metal box sitting on the top of the tower. Once this concrete had gained sufficient strength, the metal box or shutter into which the concrete had been placed was raised by hydraulic rams to create a new container into which the next lift of concrete was placed. So, incrementally, the towers grew in height, becoming monolithic structures with no physical joints. The average rate of growth of the towers at Øresund was 4m per week.

The special girders for the high bridge with the side brackets for the hanger cables were not manufactured in Spain but were fabricated in Karlskrona, Sweden. Delivered to Malmö, they too were carried to site by the lift ship *Svanen*. The high bridge girders needed to be temporarily propped until the hanger cables had been fixed to them and stressed so that they could carry the weight of the girder back to the bridge towers.

Up to 2,000 people worked on site at any one time, with around 95 per cent of the site workforce coming from the Øresund region. During the peak period of construction, around 5,000 workers were directly involved in the project, including those pre-fabricating materials off-site. On 14 August 1999 a ceremony took place as soon as the last girder had been landed on its supporting piers. With the heavy-lift crane still in position, Crown Princess Victoria of Sweden and Crown Prince Frederik of Denmark embraced symbolically at the midpoint of the bridge celebrating the physical joining together of their two countries. The official inauguration of the Øresund crossing took place on 1 July 2000 when Queen Margrethe II of Denmark joined King Carl Gustaf XVI of Sweden on the bridge, now part of everyday life for many thousands of people from Copenhagen, Malmö and the surrounding districts.

Now for a change of location. In recent years, society has become more aware of rising sea levels and, in many locations, engineers have built structures to protect cities from the worst consequences of severe storm surges. The iconic Thames Barrier in London is a typical example of such a structure. In November 2012 New York suffered a serious storm surge that caused much misery and a temporary loss of subway routes and electrical power. Civil engineers are now planning a storm-surge barrier to be built outside of that city so as to avoid future loss of key infrastructure. Venice suffers every year with floods and a storm-surge barrier is under construction outside the city to cure the problem. When complete, this barrier with its three ship entrances will enclose a large area of otherwise open sea.

Many tens of times larger than the Thames Barrier is a recently completed storm barrier across the very end of the Baltic in the Gulf of Finland to protect St Petersburg from storm surges. Known as the Venice of the north, St Petersburg, with its many river channels and canals, is a beautiful city but one that is very prone to flooding, experiencing on average one a year since its founding in 1703. Built on marshland where the delta of the River Neva meets the Gulf of Finland, the centre of the city is only a metre or two above normal sea level so is very susceptible to storm surges originating in the Baltic. In the past thirty years, flooding increased to often become a twice-yearly event, and it was constantly feared that a severe flood event could occur at any time. The city floods when levels rise to between 1.7m and 1.8m above the zero watermark for the River Neva, as recorded on a tide gauge located near the Institute

Blagoveshchensky Bridge, St Petersburg.

of Mining and Metallurgy in the centre of town. The Institute, a Neo-classical masterpiece by Andrey Voronikhin, is Russia's oldest higher education college devoted to engineering and is where Vladimir Putin studied.

Large passenger ships are unable to travel into the heart of the city, having to dock in a new deep-water port that is a bus ride out of town. Travelling on smaller ships, I have often berthed right in the centre of town at 'English Quay', located just before Blagoveshchensky Bridge, the first of many bascule bridges crossing the River Neva. The quay is a short walk along the riverside to the Hermitage Museum and other city-centre attractions. Blagoveshchensky Bridge, one of 342 bridges in St Petersburg, was built between 1843 and 1850 by Polish civil engineer Stanisław Kierbedź to replace a temporary structure, becoming the first permanent bridge built across the River Neva in St Petersburg. Substantially rebuilt between 2006 and 2008, it is illuminated at night, along with other city-centre bridges, creating a spectacular sight. All of the bascule spans along the river are opened during the night to permit the passage of large river barges navigating inland routes that can take them to Moscow and beyond.

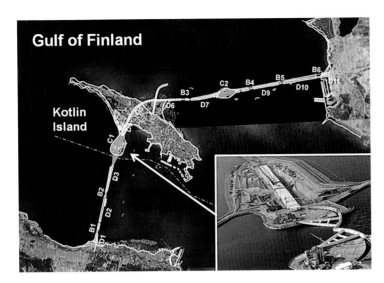

Plan of the St Petersburg storm barrier. (Royal Haskoning)

Some 290 floods have been recorded in the history of St Petersburg. The most catastrophic of all was that of November 1824. Winter came early to Russia that year; the very cold weather caused pack ice to form on the River Neva near to St Petersburg. Water that had not frozen backed up behind the ice until a spell of mild weather melted the ice jam, overwhelming the city with near freezing water, making staying alive in it for any length of time a near impossibility. Contemporary reports stated that the entire city was flooded with up to 10,000 persons having perished and much of the city's rich cultural history lost. Over 300 buildings were completely destroyed and over 3,000 were damaged, and the surge of water was apparently so powerful that ships in the harbour were also destroyed. The story of the flood was immortalised by Alexander Pushkin in his poem 'The Bronze Horseman'.

Facing the River Neva is an impressive monument commissioned by Empress Catherine the Great as a tribute to her predecessor and founder of St Petersburg, Peter the Great. Since the time of Pushkin's famous poem of 1833, the statue has been known as the Bronze Horseman, after the title of that poem. The statue is now one of the symbols of St Petersburg, in much the same way that the

Statue of Liberty is a symbol of New York City – both of them, by the way, designed and built by French civil engineers. According to a nineteenth-century legend, enemy forces will never take St Petersburg while the Bronze Horseman stands in the middle of the city. During the Second World War, the statue was not removed but was protected with sandbags and a wooden shelter, surviving the 900-day Siege of Leningrad virtually untouched.

Another catastrophic flood (3.7m above the zero water level) occurred on 23 September 1924 when 61 per cent of the city was inundated and some 600 people lost their lives. Nineteen bridges were damaged and great disruption was caused in streets paved with wooden cobbles; as the water rose the cobbles floated, causing ankle-breaking chaos for pedestrians and extreme difficulty for motor vehicles.

In 1999, the city experienced another serious flood when the Neva River rose 2.7m above normal level. The bill for damaged infrastructure came to around £14 million, and that's without counting the human and cultural costs. The projection of loss for future floods were so great that the Russian government determined that such catastrophes should not be allowed to happen, and to prevent them a storm barrier was conceived that would be built 25km west of the city centre.

Actually, it was two storm barriers: one between Kotlin Island (Kronstadt Fortress) and the mainland to the east and the other between the island and the mainland to the south. The structure was to comprise 25.4km of earth embankments topped with a six-lane motorway. UK civil engineering consultant Halcrow had overall design responsibility for the project; it secured necessary government approvals, supported the international tendering process and provided support during the construction phase. Halcrow's design aims to protect the city from storm surges with a return period of one in 1,000 years, and as such save one of Europe's greatest cities from sinking back into the marsh on which it was built.

North rotating-sector gate at the main ship channel through the St Petersburg storm barrier.

In order to cater for the normal rise and fall of the tide, and to permit fluvial flood flows from the River Neva to escape, six sluice complexes have been constructed along the length of the barrier. When open, the sluices together with two navigation channels permit the flow of water in and out of this now enclosed end of the Gulf of Finland. Each sluice complex is equipped with either ten or twelve 24m-wide steel radial gates, with sixty-four gates being provided in all. At the time of a surge, these gates and the navigation channel gates will be closed.

The 110m-wide 'secondary' navigation channel, provided through the barrier for the passage of smaller ships and barges, has been constructed to the north-east of Kotlin Island. On either side of it, a 1.5km-long concrete viaduct was built on a rising gradient to lift the motorway up and over the shipping channel. A steel vertical rising gate, weighing 2,500 tonnes, lays hidden in a slot at the base of the navigation channel. The gate is lifted to close off the channel during a surge event and can even be lifted above normal water level for maintenance access purposes. For the passage of ships with a larger air draft, the motorway bridge can be lifted to a higher level, at which time road traffic is halted.

Just to the south of Kotlin Island the 'primary' navigation channel through the barrier is located, the one through which all cruise ships and other large vessels have to navigate in order to reach the city. Engineers designed this channel 200m wide so as to minimise the adverse effects to ships' steerage caused by fast tidal flows. Designed to accommodate ships of up to 100,000 tons dead weight, the channel, in times of storm surges in the Baltic, is closed off by the swinging into action of two of the largest moving structures of their type ever built.

On each side of the channel located in a radial slot is a 4,500-ton steel rotating sector gate. The gates, each pivoted at the end of a 130m-long radial arm, are constructed like ships, designed to float to support the outer weight of the radial arms. When the sector gate is fully withdrawn into its radial slot, a pair of smaller gates can be brought into action to close off the slot, which can then be pumped dry, turning the large radial slot into a dry dock to allow maintenance to be carried out on the main gate. When the small gates themselves need maintenance, a temporary steel dam is craned into position into a preformed slot in the concrete structure, allowing the area around the small gates to be pumped dry.

Thanks to the enterprise of twenty-first-century civil engineers, St Petersburg, along with its wonderful art treasures and, of course, its residents, is now protected from future surges in the Baltic Sea and the Gulf of Finland.

11

THE RESTORATION OF THE KENNET & AVON CANAL

SAVED BY A £25 MILLION NATIONAL LOTTERY GRANT

I achieved great pleasure as Restoration Project Manager in spending millions of pounds of lottery money to save a precious inland waterway from certain oblivion. A 'Celebration in Light' was specially funded by the Heritage Lottery Fund to mark the successful completion of the project. In the UK over the past few decades there has been a major resurgence in the restoration of derelict inland waterways, and the restoration of the Kennet & Avon Canal is one of the largest so far undertaken. Essentially restored for leisure purposes, this canal, like so many others, has also proved to be an engine for the urban regeneration of the towns through which it passes.

As the name of the waterway suggests, it joins the River Kennet (which flows from the Marlborough Downs through Hungerford and Newbury to the River Thames at Reading) to the River Avon (which flows from the Cotswold Hills through the cities of Bath and Bristol to join the Severn Estuary at Avonmouth). Until the early eighteenth century, the River Kennet, between Newbury and the Thames, served only as a source of power for the many mills along its banks. Then, between 1715 and 1724, the river was canalised (made navigable) by civil engineer John Hore (c. 1690–1763). In total, he excavated 18km of artificial cuts to create a new 30km-long transport route from the River Thames right into the heart of the Berkshire market town of Newbury, where he built wharves for the loading and unloading of goods from river barges. The new navigation required the construction of twenty locks, seventeen of which were originally turf-sided; all but one of these turf-sided locks have now been replaced with brick or steel chambers. All the bridges were of timber construction, either fixed at high level or designed to swing just above water level to allow the passage of boats. The River Avon was made navigable from Bristol to Bath between 1724 and 1727, again by John Hore. He built six locks along this 18.5km-long navigation, five of which were built in short cuts alongside existing mill weirs. The waterway meant that foreign goods arriving at

A 'Celebration in Light' at Caen Hill Locks, Devizes, Wiltshire. (Robert Coles)

Bristol docks, as well as goods manufactured in Bristol, could now be conveyed cheaply and conveniently to the city of Bath and that Bath stone, used for the construction of Bristol's prestigious buildings, might be conveyed in the opposite direction.

Between 1794 and 1810, these now navigable sections of the two rivers were joined together by John Rennie, civil engineer for the Kennet & Avon Canal Company. The 90km-long broad-beam canal was provided with seventy-eight locks. Born the son of a farmer at Phantassie, East Lothian, Rennie followed a rudimentary education at the local parish school before attending higher school in Dunbar. Between 1780 and 1783, he studied engineering at the University of Edinburgh. During his career, he was to design many bridges, canals and docks. Given that the dates of construction of the Kennet & Avon Canal pre-dated the age of photography, there are sadly no photographs of the actual building of the canal. However, very early photographs of maintenance gangs working on the canal, which were taken a few decades later, do exist, and they capture the spirit of navvies working in teams with pick axes, shovels and wheelbarrows just as they would have done when building the canal for Rennie. Incidentally, the term 'navvy', used to describe the manual construction workers who built railways and roads in the nineteenth and twentieth centuries, came from the era of canal building when the gentlemen who were building navigations across the countryside were referred to as navigators, a name that was later shortened to 'navvies'.

The Canal Company had already purchased the majority shareholding of the Avon Navigation in 1796, and after the completion of the linking canal in 1810 they went on to purchase the Kennet Navigation in 1812. These three waterways, together with the River Thames, provided an inland navigation right across southern England linking the Port of Bristol and the Port of London. This route avoided ships journeying between the two cities being attacked by French privateers who, at that time, were active along

The Kennet & Avon Waterway

Bristol
Bath
Devizes
Pewsey
Hungerford
Newbury
Reading

- 87 miles Bristol to Reading (River Thames)
- 104 locks, 215 bridges

STATUTORY PROTECTION OF THE WATERWAY

- 203 Listed Structures
- 2 Areas of Outstanding Natural Beauty
- 5 Sites of Special Scientific Interest
- 24 Conservation Areas
- 1 World Heritage Site

the south coast of England. Ironically, French prisoners of war were used as part of the construction labour force to build the canal.

The distance between Bristol and Reading on the River Thames is 140km. The waterway required 106 locks to lift vessels from sea level at Bristol to the top of the Marlborough Downs and back down again to river level in Reading. Completing their journeys on the River Thames, vessels had to negotiate another twenty-one locks to reach sea level in London.

Following the passage of the last commercial vessel, the canal lay derelict and unloved for over fifty years, until it was slowly reawakened by the efforts of volunteers who raised small sums of money, over a very long period, to start a piecemeal but worthwhile restoration.

Given the statutory protection measures listed, you might well imagine that project-managing the restoration of this waterway was a nightmare, and you would be absolutely right! So, why bother to restore this derelict canal that was originally built as an industrial trading route for horse-drawn boats – a route that before the coming of the Great Western Railway (and, later, the M4 motorway) carried most of the commercial traffic between London and the

west of England. Well, for one reason, the canal is still being used for horse-drawn boats, but now carrying a human cargo, enjoying what must be the most relaxing way to travel through the English countryside. But today, canals as a leisure resource are not just for boats. The towpaths provide wide and level footpaths offering access for all and they are also used as cycle routes. These paths are used for jogging by some and as a tranquil place by others where they can just amble and stop and stare at the antics of amateur navigators in the locks. There is actually a word to describe such groups of onlookers – 'Gongoozlers'. On every waterway, anglers can be seen enjoying their sport, while groups of school children are often seen making use of what is a tremendous outdoor educational resource.

The fifty years of hard work undertaken by the volunteers to bring the waterway back to life culminated in the completed canal being opened by Her Majesty Queen Elizabeth II in 1990. Unfortunately, through-navigation was to be short-lived due to the deteriorated state of major structures along the waterway, structures that had simply been beyond the capability of volunteers to finance and to repair.

The case for the full restoration of the Kennet & Avon Canal having been made, a partnership was put together comprising the voluntary Kennet & Avon Canal Trust, fifty-one local businesses, seven local authorities and British Waterways. Once established, the partnership submitted an application to the Heritage Lottery Fund for a capital grant to complete the earlier work started by those volunteers. After many detailed presentations by the partnership, a grant of £25 million was made from National Lottery funds towards the project cost, which had been estimated at £30 million. This lottery grant was (and for some time afterwards remained) the largest lottery award ever made to a single project.

Despite this generous grant, £6 million had to be raised as 'matched funding' by the partnership. Well-known personalities such as husband and wife actors Prunella Scales and Timothy West, who are also both vice presidents of the Kennet & Avon Canal

Author's steam launch *Scathtan*. (Doug Small)

Caen Hill Locks, Devizes, Wiltshire. (Robert Coles)

Trust, joined in this fundraising effort. Ordinary members of the Trust devised many ingenious ways of raising matched funding. I ran boat rides in my coal-fired steam launch – this was at a time when James Cameron's *Titanic* was playing in cinemas. Although explaining that the boat's steam engine was a miniature version of those that drove *Titanic*, the poster advertising the trips promised that there would be no icebergs on the Kennet & Avon Canal!

Before any engineering design work, let alone construction, was undertaken, the lottery authority demanded that we prepare a conservation plan. The plan looked at built and natural heritage, at the operational history of the canal and at best practice in the leisure use of waterways. The document started with a statement that the restoration project was designed to: create a sustainable canal that would stand up to twenty-first-century use; preserve, restore and recreate wildlife habitats; maintain its original built and natural heritage; and enhance the canal for visitors, local communities and businesses. The production of the plan was quite a complicated process. First, many surveys were undertaken and these were fed into a draft conservation plan. An example of the type of detailed surveys undertaken was that into the existing natural environment – surveys of the water column, canal banks, verges and hedgerows

identified key habitats and species. The draft plan was then worked up into a final document, taking into account current legislation and local authority planning policies. Once completed and approved by the project monitors, the conservation plan could then begin to inform the design process and, subsequently, the construction operations. Regular reviews were undertaken and relevant findings were fed back into the design process through design team meetings. Over a longer period, sustainability was monitored to inform management decisions in the maintenance field and, finally, an information exchange was established to inform follow-on projects of our experience. The conservation plan was probably the most important document to come out of the project and was, for many years, held up by Heritage Lottery as the example that all future waterway restoration projects should follow. Copies of the plan were deposited in the libraries of learned institutions and public lending libraries along the route of the waterway.

Four project monitoring organisations were appointed by the Heritage Lottery Fund: English Nature; The Countryside Agency; English Heritage; and the international civil engineering consultant Arup, who was appointed as lead monitor. As each of these national bodies had its own ideas about what should or should not be included in the project, and as each regularly expressed its own views on how every separate element of the project should be managed, their combined input presented an interesting challenge for the project manager.

The first major problem with which we had to deal was canal leakage, much of which stemmed from the long period of dereliction when the original clay-puddle waterproof lining of the canal had dried and cracked, allowing the water that remained to escape into the surrounding ground. We devised many individual solutions to tackle this problem but all of them required the complete draining down of the section to be worked. An essential first operation before any section of the canal was drained was to rescue the fish. This was achieved by passing an electric current through the water, a process known as electrofishing. The hundreds of stunned fish resulting from this process then lay on the surface of the water, from where they were collected in nets and transferred to another part of the waterway. Perhaps, surprisingly, the fish mortality rate from this operation was very low.

At Caen Hill, near Devizes in Wiltshire, John Rennie had provided sixteen locks that had been constructed close together, each with its own reservoir to store water for lock operation. On the average run of a canal, locks are spaced approximately 0.8km to 1.5km (½ to 1 mile) apart so there is sufficient water in the pound between the locks for their operation, but this was not the case at Caen Hill, hence the provision of the reservoirs or side ponds, several of which leaked excessively through their beds. Because of the listed status of the lock flight (it enjoys the same degree of protection as Stonehenge) we were permitted by our own conservation plan to use only traditional lining materials. After the ponds were drained down and all soft silty material removed, a new lining of puddled clay was worked into the surface. Rennie originally specified a 1.2m thickness of clay lining, which was put down in layers and 'puddled' by driving flocks of sheep backwards and forwards across the surface, thereby working the clay into an impermeable membrane. We didn't use flocks of sheep to puddle the new clay lining but, instead, used a piece of civil engineering machinery appropriately called a 'sheep's foot roller', built just like a road roller but, instead of a smooth drum, it has one encrusted with knobbly projections, therefore having the same effect as a flock of rather heavy sheep. One thing is certain, however, the navvies who built the canal for Rennie would not have appreciated such a machine, it being recorded in company minutes that those navvies often liberated parts of their 'puddling machine' to provide a well-earned Sunday meal.

Following the relining of the side ponds with clay, the flora quickly recovered because we had taken great care to remove the emergent plants from the edges of the reservoirs before any excavation work was undertaken. The plants were then kept in purpose-built storage lagoons throughout the winter while the works were in progress and in the spring they were in excellent condition to be replanted around the edges of the reservoirs. On completion of the work, the flight of twenty-nine locks at Devizes (of which the Caen Hill flight of sixteen locks is the centre section, and one of the Seven Wonders of the Waterways) held water and operated as originally designed, lifting vessels nearly 300ft from the lower Wiltshire plain up to the level of the Marlborough Downs.

Another problem with which we had to deal was bank erosion. Remember that canals were designed for horse-drawn boats, whereas vessels today have propellers, the turning of which washes the edges of the canal away. Standard maintenance practice over many years was to drive steel sheet piles along the edge of the waterway, creating a rather unsightly and austere wall of steel. Apart from its appearance, it is not at all inviting to protected species such as the water vole, requiring easy access to the natural bank. Our intention was to avoid using the steel sheet pile solution whenever it was sensible to do so.

Typical steel pile canal edge reinforcement, here in use on the Langollen Canal in North Wales.

We actually developed a variation of the steel pile theme by hiding the pile below the surface of the water. A timber fender was fixed to the front face of the pile in order to protect boats that might inadvertently come into contact with it. The pile was backed with a roll of coir, a fibrous coconut material, which was then planted with emergent plants. Access to the clay bank of the canal for water voles and other amphibious creatures was then possible over the tops of the piles and coir roll. Incidentally, the water level of the canal in the photograph opposite had been temporarily lowered to assist planting and, therefore, the coir and the timber fender are visible. Both would be hidden once the water had been returned to its normal level.

Another development of this 'reinforced soft edging' approach was our use of woven hazel faggots, which, in canal terms, we

Hidden steel edge piling at Sells Green, Kennet & Avon Canal.

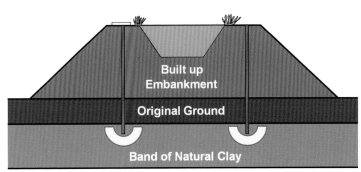

Cross section through a typical embankment illustrating 'containment piling'.

believe was a first. The system was a great success. Water voles loved this solution, as it provided them with easy access to the clay banks into which they could burrow. Plants and young fish were protected from boat wash and thrived, and even the boating community liked this edging because it was very forgiving to their vessels when they came into contact with it.

A hard edge for boat moorings, for the safe boarding of vessels, had to be provided at some locations. But instead of a continuous hard edge, particularly in areas of valuable natural habitat, we developed a 'hit and miss' solution, which we termed a 'castellated mooring'. A natural edge to maintain access to vole-friendly habitat was interspersed between hard-edge sections.

In many places along its length, the Kennet & Avon Canal is carried on embankments. Leakage in these areas can lead to a breach – the most catastrophic of all canal failures! Breaches are born when a slow leak through an embankment is allowed to run unchecked, or perhaps even goes unnoticed. The resulting 'piping' failure gets bigger and bigger until the embankment becomes unstable and a whole section is carried away. Canal water cascading through such a gap removes more and more of the embankment together with much of the canal

bed, allowing all of the water within a given canal pound (the stretch of water between two locks) to inundate adjacent property.

Where the local geology permitted – for example, where a band of natural clay lay reasonably close to the ground surface – our preventative design solution was to drive steel sheet piles through the heart of the embankment and into that impermeable clay layer. A waterproof trough, with steel sides and a clay base, was thereby created to contain the canal water safely.

Martinslade Embankment, near Devizes, was the first embankment where we employed this technique. Running diagonally across the centre of the picture on page 166, from bottom left to top right, the embankment was leaking canal water into the marshy area between the canal and a railway embankment just visible on the extreme right. Steel piles can be seen laid out along the top of the embankment, prior to being driven into the heart of the embankment by a floating piling rig. Since the piles had not been driven along the edge of the waterway, but instead had been driven along the centre line of the bank, just one year later the natural waterway edge had recovered with no visible signs of the steel piles that lay buried beneath the restored towpath. Superb water vole habitat

Brown Rat
SHARP POINTED NOSE
Big ears
Long sleek body
Weighs about 500g
Long thick, furless tail.

Water Vole
ROUND BLUNT NOSE
Small ears
Shorter much rounder body
Smaller, about 200g to 350g
Shorter thinner fury tail.

Above: The principal differences between the common brown rat and the endangered water vole.

Left: Steel piling at Martinslade Embankment, near Devizes. (British Waterways)

had been maintained and the appearance of the embankment, as viewed both from passing boats and by walkers enjoying this beautiful countryside, could not be bettered.

Water voles are a protected species under the European Habitat Directive and, today, the Kennet & Avon Canal is probably one of the better water vole habitats in the south of England. One has, therefore, to be very unlucky when travelling along the canal not to see one. The major difference between this endangered species and the common brown rat (also to be seen swimming in canals) is that the water vole has a distinctively blunt nose, whereas the rat has a long, pointed nose. (It was once suggested to me that many of the sightings of water vole on the Kennet & Avon Canal were actually pointed-nose brown rats that, when scurrying through their burrows, had accidentally discovered my buried steel pilings!)

At Sells Green, also near Devizes, the canal is carried on another embankment that leaked badly and often flooded the lower lying ends of the adjacent fields as they gently sloped up the hill away from the canal. Initially, a piled solution similar to that employed at

Martinslade was thought to be the solution. However, after engaging in a bit of lateral thinking, we decided to purchase the parts of the fields that flooded and simply allow them to flood, thus creating a sizeable lake at the same level as the canal. This new lake became a win–win element of the project. The land required for the lake was owned by a church in the nearby market town of Devizes, a church that desperately needed a new roof. The money we paid for the land provided that a new roof and the cost of the land for the lake was no more than we would have spent on steel piles. The new lake was then developed as an attractive wetland nature reserve, with its flooded contour becoming a shallow beach and openings between canal and lake creating island refuges for wildlife – openings that were closed off with wire mesh to prevent fish entering the lake. With low numbers of fish in the lake, water fleas or Daphnia (voracious eaters of algae) proliferated and cleared the water, allowing plants to thrive. Sells Green Lake is now an attractive area of varied habitat complete with new hedgerows and purpose-built sandy clay banks used by kingfishers to construct their nesting burrows. I am

Crofton Pumping Station, near Great Bedwyn in Wiltshire, with a restored Great Western Railway King-class locomotive passing by. (Robert Coles)

sure that to this day the lottery authority had no idea that its 'canal' grant had also paid for a new church roof.

Because the canal channel had become badly silted during the decades of dereliction, dredging of the navigation was required over tens of kilometres of its length in order to re-establish the original profile cut by Rennie. He designed the main channel at 1.4m deep but designed the short summit pound, at the canal's highest point, at 2.4m deep to act as a reservoir of water for the operation of the locks down both sides of the hill. Several dredging contractors were employed and fields adjacent to the canal were rented from local farmers. Once approved by the Environment Agency as being of beneficial quality to agriculture, the dredged material was spread thinly over these fields. Just one year later, the farmland was back in full production. The material dredged from the canal had effectively been returned from whence it came before the action of rainfall over several decades had washed it into the canal.

Despite the 2.4m depth of the summit pound, shortage of water to operate the locks, particularly in the summer months, was always a problem for the Kennet & Avon Canal. On the top of the Marlborough Downs there is not even a small stream that might feed water into the canal. Rennie's solution was to construct both a reservoir further down the hillside (where it could collect rain run-off) and a pumping station to lift water from the reservoir up to the canal.

Having purchased this facility from British Waterways some time before the start of my project, Kennet & Avon Canal Trust volunteers had already restored the engine house and the steam pumps to a fully operational condition. The 'newer' of the two beam engines, an 1846 machine from Harveys of Hayle in Cornwall, shares the engine house with the granddaddy of all operating beam engines, an 1812 Boulton & Watt engine. The *Guinness Book of Records* once recorded that this engine was the oldest anywhere in the world still doing the job for which it was designed and still on its original site.

The task facing my project was to rebuild the chimney, which in 1958 had been reduced to half its original height, having become unstable. An uncanny sensation was experienced by all of us on top of the chimney on the day that we celebrated the topping out of the completed work. The scaffolding on which we were standing, which surrounded the top of the chimney, gently moved in the moderate breeze. We did not sense any movement, but observing the opening and closing gap between the scaffolding and the chimney we wrongly gained the impression that it was the substantial brick chimney that was actually swaying. The chimney builder, perhaps concerned at what people might think of the quality of his bricklaying skills, looked a little worried but, of course, need not have done so. The Crofton Pumping Station is well worth a visit at any time of the year, but particularly on UK bank holidays when both engines are steam operated. Another treat for the steam buff is that the West of England main railway line runs alongside the canal at this point, actually squeezing between the pumping station and the canal, and this line is often used by steam-hauled special trains.

The other significant pumping station built by Rennie to provide water for the canal, and which has also been restored to working order by volunteers, is at Claverton, some 8km south of the city of Bath. At this location, the River Avon flows parallel with the canal, which is located some 12m above the river and on the other side of a railway. The renewal of the cast-iron pipeline from the pumping station, through a tunnel under the railway and up the hill to the canal, was the contribution our project made to the continuing operation of what one might describe today as Rennie's environmental masterpiece.

Claverton pump is, to my mind, the ultimate 'green machine'. The flow of the River Avon turns a huge water wheel of the size used to propel a Mississippi steamboat. The wheel turns a wooden-toothed gear train, which rotates a cast-iron flywheel to which cranks are connected. These rotating cranks move a pair of iron beams up and down. To the other ends of these rocking beams are connected rods

Claverton Pumping Station, near Bath. (Robert Coles)

that activate pumps that draw water from the river to deliver to the canal. No external or internal combustion, just the energy from the river forcing some of its own water 12m up to the canal.

With water being such a precious commodity on the Kennet & Avon, we set about improving its conservation. Many lock gates that leaked badly were replaced with new hardwood gates, the timber for which had to be verified as having been obtained from a sustainable source. We built by-weirs in the pounds between the locks to accurately set the water level of the pound. Water flowing over these small weirs was piped to the next lower pound and so, instead of being lost by overtopping banks and being carried away from the canal by local ditches and streams, the precious water was retained within the canal itself. To avoid whole lockfuls of water being wasted downhill every time a lock is used, we built back-pumping stations below each flight of locks. These stations contained submersible electric pumps designed to return some 80 per cent of the water used in lock operations on the flight back to the top of the flight from where it could be used again.

My project included significant work to both Dundas and Avoncliffe aqueducts, the largest structures of their type on the

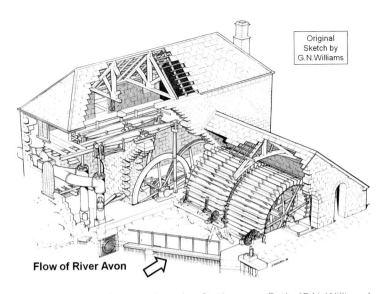

Flow of River Avon

The anatomy of Claverton Pumping Station, near Bath. (G.N. Williams)

Dundas Aqueduct on the Kennet & Avon Canal. (Robert Coles)

waterway, both of which had been given full classical treatment by Rennie. From Bath, heading south, Rennie had built his canal on the west side of the River Avon. Just south of Dundas, however, the wide Midford Brook joins the River Avon and, a little further south, so does the River Frome. As the confluences of both of these tributaries are on the west side of the river, to avoid them and their associated low topography, Rennie crossed his canal to the east side of the River Avon, bringing the canal across the river on the Dundas Aqueduct and then, 5km further on, crossing it back again to the west side on the Avoncliff Aqueduct.

Dundas Aqueduct is an engineer's version of a Roman triumphal arch. There are few aqueducts anywhere with such a level of grandeur. Dundas Aqueduct is built of Jurassic limestone (known locally as Bath stone) and is a Scheduled Ancient Monument and a Grade I listed building. The condition of the structure at the outset of my project was poor. Large areas of the stonework had deteriorated and, in many places, had been replaced by engineering bricks. These repairs date from the time that the waterway was managed by the Great Western Railway Company, who had purchased the canal in 1852. I recall long philosophical discussions with English Heritage,

who wanted the brickwork to remain in place because, in their view, it contributed to the story of the life of the structure. Our view, however, prevailed, after I convinced them that water migrating through the soft limestone was being trapped above the non-porous engineering bricks, thereby damaging the overlying stone. Deterioration was particularly rapid during frosty conditions when the trapped water would freeze, causing the stone to spall away. Another interesting discussion concerning this aqueduct was held, this time with the two local planning authorities in whose territory the structure sits. The boundary between them followed the centre line of the river and so each authority was responsible, in planning terms, for one half of the structure. I wanted to replace the severely damaged Bath stone copings at the edge of the channel with concrete blocks made from an aggregate of the same colour as Bath stone and using coloured cement, believing that the blocks would look like Bath stone but would be more robust for modern-day use, i.e. being regularly impacted by steel-hulled vessels. I lost this particular battle. Whilst West Wiltshire District Council approved the proposal, Bath and North East Somerset District Council stated that as Bath was a World Heritage Site then only Bath stone would be permitted.

Avoncliff Aqueduct on the Kennet & Avon Canal.

The sag in the structure of Avoncliff Aqueduct on the Kennet & Avon Canal.

My argument even quoted Rennie's report to the Kennet & Avon management committee of 25 March 1803, where he implored the committee to allow him to use bricks in areas of working impact because of 'the badness of the local stone'.

Avoncliff Aqueduct was of a very different design; its main arch is more elliptical in shape than the main arch at Dundas and, as captured in the photographs above, is a little misshapen. If a string line is applied along the top, the whole structure appears to sag in the middle. The stone arch was built, as all arched structures have been down through the ages, on temporary timber centring. Once completed and closed with a keystone, and having gained its full strength, then the temporary timber support would have been removed, at which point it is believed the Avoncliff main arch relaxed somewhat.

The dip in the masonry is very evident when viewed along the towpath. Despite any early movement, the structure had been standing safely for some 200 years and so we did nothing other than remove engineering bricks and replace or reface many of the stone blocks in the arches and elsewhere. Hieroglyphic-style marks on the original stone blocks were identified as masons' marks. Individual masons applied their unique mark to stones that they had placed, these later being counted to determine payment due for their work. Paying homage to this historic practice, I encouraged the masons working with us to apply their own marks to the replacement stones.

Bruce Tunnel is the longest of three on the waterway. The tow-path did not go through the tunnel; instead, towing horses were led on a path over the top of the hill. Fixed to the brick lining in the eerie darkness of the tunnel are the remains of chains used to pull boats through. At the time of its building, a large commemorative stone tablet was placed above the east end portal by the Kennet & Avon Canal Company formally 'inscribing the tunnel with the name Bruce' in recognition of the support given to the project by Charles Lord Bruce and his father the Earl of Aylesbury. That, at least, was the originally intended message but, at the time of my project, the inscription cut into the decaying stonework was almost unreadable. English Heritage, probably quite rightly, did not want this historic plaque to be altered in any way. Having discovered that the family firm of stone masons who had produced the original tablet was still trading (and from the same yard in the Wiltshire village of Great Bedwyn), I proposed to English Heritage that a really interesting

Bruce Tunnel's eastern portal, with the original commemorative stone tablet above. (British Waterways)

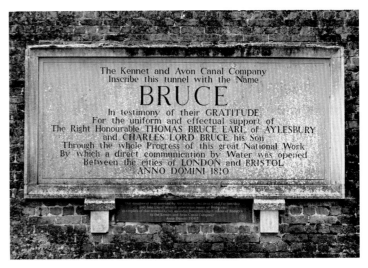

Replica tablet at Bruce Tunnel with an interpretation panel beneath.

story might be told if a smaller replica tablet produced by this historic family firm could be placed alongside the tunnel mouth.

English Heritage supported the proposal and beneath the replica tablet, interpretation has been provided linking John Lloyd, stonemason of Bedwyn to the Restoration Partnership, to his seventh-generation ancestor Beniamin Lloyd, stonemason of Bedwyn to the Kennet & Avon Canal Company.

To the east of the city of Bath, a considerable length of the canal not only leaked badly but the local geology meant that whole sections of hillside into which the canal had been cut were relatively unstable. Rennie recognised this and provided emergency self-closing gates at regular intervals along the waterway. All of these gates were thought to have been removed over time, as none were known to exist until, while excavating an engineered narrowing of the canal, we found one buried deep in the canal bed and in surprisingly good condition. My plan was to exhibit the gate on brick plinths on the flat area of ground on the off side of the narrowing where it was found, along with an explanation of the reason why the narrowing was constructed in the first place. The plinths would have exactly mirrored those plinths in the canal bed that we

had discovered below the buried gate. My intention was to provide interpretation on the towing path side explaining the unique way that Rennie had designed these gates to operate. For reasons that I still do not understand, English Heritage thought this not a good idea, insisting instead that the gate be sawn in half and displayed in the canal museum in Devizes.

Throughout these sensitive lengths of the waterway and after much preparation work, including, with the permission of English Nature, the capture of the resident water vole population and packing them off on their holidays, we completely rebuilt the canal structure. Having removed all of the silt from the channel, a stone haulage road was constructed in the bed of the canal to provide access for construction vehicles. From this road, the sides of the canal were profiled, complete with provision for planting shelves, and new surfaces were pitched with stone and blinded with a sandy grit. A continuously welded PVC sheet, protected on its underside by a geotextile fabric, was then overlaid on top of the grit, similar to the way that one might build a garden pond but on a very much larger scale. After a second layer of geotextile had been laid on top of the PVC membrane, profile boards were placed at regular intervals

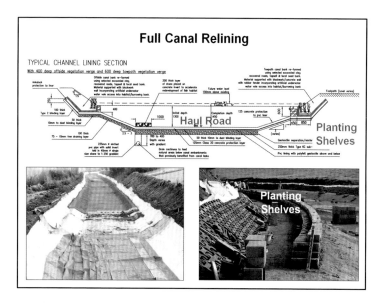

Detail of the full canal relining on the Kennet & Avon Canal near Bath.

Author with the Kennet & Avon Canal Trust chairman, David Lamb (on the right), after the 'unofficial' opening ceremony of the restored canal. (Robert Coles)

along the channel. Between these boards 5in of concrete was then placed using a mobile pump to deliver the concrete through a temporary pipe running down the centre line of the channel. This concrete was to protect the lining from mechanical damage. Concrete block walls were built on the outer edge of the planting shelves and, as these would end up below water level, they were faced with a hard rubber fender to deal with possible boat impact.

Care was taken in returning the original waterside plants to their new home on the planting shelves. Just like at Caen Hill, these plants had been over-wintered in temporary lagoons while the rebuilding work was in progress. The new concrete bottom of the canal was 'seeded' with gravel and silt material that had been removed early on in the excavation process. All of this attention to detail immediately paid off just one season later and the results were astounding. People have commented that the waterway no longer looks like a man-made canal but much more like a natural river.

Finally the water voles were returned to the canal from their temporary 'holiday homes' and I am pleased to say that they are now really well established in their new engineered habitat. The

project was the recipient of many national awards for restoration and innovation, one of which I am particularly proud of was presented to me by The Engineering Council for 'Engineering in the Natural Environment'.

On a rainy day in May 2003 the waterway was formally reopened by HRH The Prince of Wales. However, the weekend before the royal reopening, I was asked by the Canal Trust (in recognition of my six years of leading the team managing this phenomenal restoration) to cut a ribbon across the canal at Devizes and to lead, with my steam launch *Scathtan*, a flotilla of some forty narrow boats through the town to declare the waterway 'unofficially' open – what an honour!

The Kennet & Avon Canal, due in big part to the Heritage Lottery Fund, once more links by water the historic ports of London and Bristol. It provides great enjoyment for people of all ages, an improvement to natural habitat and a boost to the local economies of the communities through which it passes.

12

THE IMPOSSIBLE DREAM

BRITAIN'S MOST AMBITIOUS ONGOING CANAL RESTORATION PROJECT

During the past few decades in the UK there has been a major resurgence in the restoration of derelict inland waterways. The Wilts & Berks Canal was completed in 1810 to provide an inland navigation between Bristol and Oxford and onwards into the industrial East Midlands. This story recounts its history, construction, dereliction and the current restoration project – a project that is probably now just about at the end of the beginning of the restoration process. It is one with which I have been personally involved for well over a decade, seven of those years as chairman and, latterly, as vice president of the Charitable Volunteer Trust undertaking the restoration work.

As well as providing a new transport link across the country, the building of the waterway achieved another improvement to the carriage of goods in the area. At the time of its construction, vessels en route from the West Midlands to London via the Stroudwater Navigation and the Thames & Severn Canal had to navigate the notoriously difficult upper Thames. This part of the river was narrow, twisty and shallow; it was furnished not with chamber locks, but with flash locks. The completion of the Wilts & Berks and North Wilts canals provided a useful bypass to this difficult area.

The Wilts & Berks Canal restoration is the longest length (and perhaps the most ambitious) of all restorations ever attempted in the UK. Much of the channel has been filled in and many of the structures have either been destroyed or have fallen into serious decay. The trust, with its 2,500 members, certainly has an exciting challenge before it. So, why bother to restore this derelict canal, which was originally built as an industrial trading route, allowing horse-drawn boats to convey large quantities of cargo that had previously been carried in small volume on pack horses or in small horse-drawn wagons?

Well, for one reason, canals can still be used for horse-drawn boats. A horse-drawn trip boat operating at Llangollen on the canal of that name in North Wales is extremely popular in the summer

A party of children pose at the former 'Lion Bridge' (an early example of a lift bridge), Swindon. (Doug Small's collection)

tourist season. We hope, one day, to attract such an operation onto the Wilts & Berks Canal, but, today, canals as a leisure resource are not just for boats, as the completion of the restoration of the Kennet & Avon Canal has demonstrated. The towpaths, as well as providing wide and level footpaths offering access for all, are also used as safe off-road cycle tracks. 'Gongoozlers' can just stop and stare at the antics of amateur navigators in the locks, and increasing numbers of schools with canals on their doorstep are discovering that these waterways are ribbons of social history and a tapestry of the natural world.

The Wilts & Berks Canal story started at Wootton Bassett Town Hall in 1793. Following a notice placed in the *Bath Chronicle* by the Earl of Peterborough, a meeting of interested parties met together to appoint a committee of management who subsequently engaged Robert and William Whitworth as their civil engineers. (Sadly, in more recent years, Wootton Bassett, and in particular its town hall, featured widely in worldwide media reports as the place where the residents of this small Wiltshire market town lined the streets in silent respect as the bodies of repatriated warriors of the Afghanistan conflict passed through the town, having been flown home to the nearby Royal Air Force base at Lyneham. In recognition of this, Queen Elizabeth bestowed upon the town the rare honour of the title Royal, and thus it became Royal Wootton Bassett.) On 30 April 1795, an Act of Parliament was passed to authorise construction of the Wilts & Berks Canal. Incidentally, the waterway has never officially been known as the Wiltshire & Berkshire Canal, due to the sloppy drafting of the Act of Parliament, during which the shortened terms for the two counties used in the draft documents were copied into the Enabling Act. In the year following the Act of Parliament, a company was formed to build and operate the canal.

The main line of this canal ran from a junction with the Kennet & Avon Canal near the Wiltshire market town of Melksham (located east of the city of Bath) north-eastwards through what was to become the railway town of Swindon to join the River Thames at Abingdon, just south of Oxford. Four short branch canals connected the towns of Chippenham, Calne and Wantage and the village of Longcot to the main line. The independently constructed North Wilts Canal linked Swindon with Cricklade on the Thames & Severn Canal. The North Wilts Canal Company was taken into Wilts & Berks ownership just one year after independent operation, thus providing the Wilts & Berks with a link from its main line to Gloucester and the West Midlands. In its heyday, its cargoes included coal transported from the Somerset coalfield. Coal barges started their journey on the Somersetshire Coal Canal and then transited the western end of the Kennet & Avon Canal before entering the Wilts & Berks Canal at Semington Junction. After leaving the Wilts & Berks at Cricklade, they continued their journey to West Midlands factories via the River Severn or, by leaving the Wilts & Berks at Abingdon, they continued to East Midlands factories via the River Thames and the Oxford Canal.

With the centre of its operation based at Swindon, the Wilts & Berks Canal Company's fortunes reached a peak during the period that Isambard Kingdom Brunel was building the Great Western Railway. His railway was also centred on Swindon and the canal carried large quantities of construction materials for the building of the 'iron road'. This prosperity was, however, short-lived, when nearly all of its trade, like that on so many other canals across the UK and around the world, was captured in the mid 1800s by the then expanding railway companies. By the turn of the last century, the waterway was virtually derelict when, in 1901, Stanley Aqueduct, a structure carrying the main line over a local river, partly collapsed; what little trade there was literally disappeared overnight. In 1914 the canal was officially abandoned by an Act of Parliament and ownership passed to riparian local authorities and adjacent landowners. Some of these new owners filled in lengths of the waterway and demolished many of the structures.

Horse-drawn passenger trip boat on the Llangollen Canal, North Wales.

Map of the Wilts & Berks Canal and connecting waterways. (John Minns)

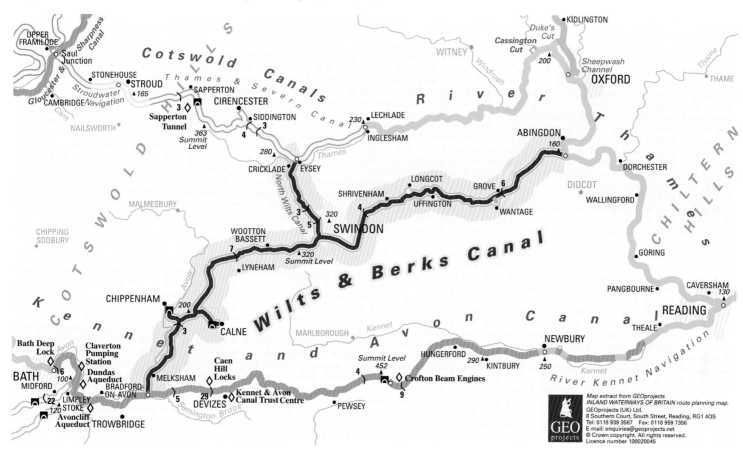

That so easily could have been the end of the story, but, in 1969, author and canal historian Jack Dalby started research into the line of the canal and two years later he published his book *The Wilts & Berks Canal*. Intrigued by what Dalby had uncovered, a small group of enthusiasts formed the Wilts & Berks Canal Amenity Group in 1977. The stated aim of the group was to trace and record what remained of this once important waterway. Examples of what they found included bridges at various stages of collapse, aqueducts no

longer able to carry the canal over rivers, and lock chambers with trees growing through the brickwork. At many locations, interesting items of early ironmongery were uncovered.

Ten years later, in 1987, the Amenity Group changed its aims to not only preserve what remained but to actively start a programme of restoration. The then impossible dream was to restore the entire navigation from end to end. As the twenty-first century dawned, the Amenity Group reformed itself to become the Wilts & Berks Canal

The former entrance to the Wilts & Berks Canal from the River Thames at Abingdon, Oxfordshire. (Doug Small's collection)

Trust. Fortunately, today, the Trust does not act alone; its vision of full restoration is shared by a strong supporting partnership under the leadership of the Marquis of Lansdowne, a well-known leisure entrepreneur. The Wiltshire, Swindon and Oxfordshire Canal Partnership is composed of all of the local authorities and town councils along the line of the canal, national organisations such as the Nationwide Building Society, the Canal & River Trust (formerly British Waterways), the Environment Agency, leisure groups such as Sustrans and Canoe England, local businesses, educational establishments and neighbouring canal trusts. The partnership strongly believes in the business opportunities that will grow from creating a sustainable leisure resource for the twenty-first century – a resource that will link to the River Thames, the Kennet & Avon Canal and the Cotswold Canals, extending long-distance footpaths, cycle tracks and bridleways and providing new access routes for local communities and visitors. It wants to preserve, restore and create new wildlife habitats and to bring back to life its original built heritage to celebrate its

history as a living memorial to those who toiled with pickaxes and wheelbarrows to construct it. In particular, endangered species such as the great crested newt and the water vole will greatly benefit from the creation of over 80km of new standing-water habitat, the fastest disappearing habitat type in the British countryside.

In 2004, the restoration proposals for the Wilts & Berks Canal received a terrific boost when the British Waterways report 'Waterways 2025' officially recognised those proposals as a serious contender in the national restoration league table. This was probably one of the most important milestones in the Trust's history. However there were (and still are) many problems to be resolved, including four significant blockages of the historic route. Development and growth of the towns of Melksham, Swindon, Abingdon and Cricklade have permanently blocked the original route, which has required the commissioning of a number of consultant studies to determine alternative routes around these blockages and to seek out new sources of water for lock operations.

Expansion of the town of Abingdon had swallowed up the old junction, the wharves and the first kilometres of canal. The original junction with the Thames was delightfully captured in a contemporary painting (above) – the canal channel off to the left, under the stone bridge, has been filled in and built over. Although Abingdon church and some of the wharfside buildings still survive, this bridge and canal have long since gone. Still surviving today, however, is an original cast-iron bridge spanning the River Ock at its confluence with the Thames. That bridge, which was built to provide access for horse-drawn wagons servicing the canal warehouses at Abingdon, proudly bears across the entire length of its arch the inscription, 'Erected by the Wilts & Berks Canal Company AD 1824 – Cast at Acramans Bristol.'

To celebrate its Diamond Jubilee in 2006, the national Inland Waterways Association set up a competition and invited applicants to bid for funding towards a suitable restoration project. We at the

Jubilee Junction – the new Thames Junction at Abingdon, Oxfordshire. (John Minns)

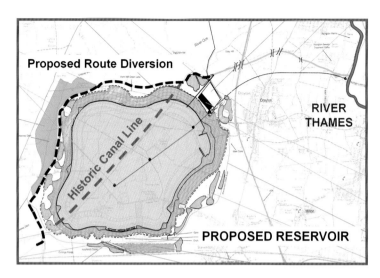

Thames Water's proposed reservoir at Abingdon, Oxfordshire. (Author; Thames Water)

Wilts & Berks Canal Trust submitted a scheme to build a new junction with the River Thames at Abingdon, together with a new length of canal to skirt the southern part of the town centre. We won that competition and received their grant. Following on from the good practice established with the Kennet & Avon Canal restoration project, the Trust first produced a basic conservation plan to inform the detailed designs for the new junction. Agreement was reached with the Environment Agency, who are the navigation authority on the River Thames and who, as a member of the Canal Partnership, already supported the project in principle. After careful consideration by each of their multi-discipline divisions, the agency later went on to give formal approval to our detailed plans.

The new junction was constructed using a successful mix of building contractors and volunteers, and on 30 August 2006 Jubilee Junction was born. It proved an immediate hit with passing Thames boating traffic, all wanting to navigate onto the Wilts & Berks Canal and to purchase the celebratory brass plaque to affix to their boats to prove that they had been there on this historic occasion. Walkers on the Thames path also took time out to have a look at this new diversion. It is just a short length at this time, with the new canal cut through into a worked out and flooded gravel pit, but the blockage at Abingdon has been bypassed and we are on our way westward towards Swindon, Bath and Bristol.

Immediately west of Abingdon, at Drayton, is a potential fifth permanent blockage to the restoration of the canal along its historic alignment. The local water authority, Thames Water, have outlined plans to build a new reservoir on that alignment, although it is by no means certain that the reservoir will be built, as their proposals are the subject of ongoing environmental and planning considerations. If the reservoir is built then the intention is to fill it during winter months with water pumped from the Thames through a large diameter pumping main. Stored water would then be returned to the river in order to supplement low summer flows through that

same pipeline. Building on the historic alignment at this location would not necessarily be all bad news because, as the route of the canal is protected within the adopted local authority plan, then the developer, in this case Thames Water, would have to make provisions for a diversionary channel. Such a new channel could well be a better starting point for restoration volunteers than the historic channel is at this location. Of possibly even greater value to the Trust, arising from Thames Water's proposals, comes a provision of the Reservoirs Act. Because the proposed reservoir is a totally embanked structure (in other words one contained by a dam that runs right around the lake) then the Reservoirs Act, which governs such structures, demands that provision for emergency drawdown of the lake is provided; however, the pipeline used to fill the reservoir would not satisfy this requirement – an open emergency drawdown channel linking the lake directly back to the Thames would be required. Such a channel would have to pass under the A34 trunk road and under several other highways. Because of all of the uncertainties created by their reservoir project, the Trust has secured an agreement from Thames Water that if they build these structures then they will all be built with sufficient headroom for

Building a new brick-arch bridge using traditional construction methods at Steppingstone Lane, near Shrivenham, Oxfordshire. (Doug Small)

navigation. The principal frustration today is not knowing which route will be followed, therefore no meaningful restoration is possible at this time anywhere along this length. If the reservoir does not go ahead, then locks that might otherwise be lost forever under tens of metres of water will be restored as progress is made along the original protected route.

Between Abingdon and Swindon many individual restoration projects have been undertaken by volunteers from the local branches of the Trust. In the centre of Wantage, at the end of the original short branch canal that linked the town to the main line,

there used to be a busy canal basin, but this has been filled in and a new housing development built over it. Surviving in the middle of these new homes is Wharfinger's House, the home of the former canal basin manager. On the edge of the development, one other historic building survives – the Sack House – which was externally restored by the housing developer and sold to the Trust for £1. Volunteers have since laid a new concrete floor, established water and electrical supplies and fitted built-in furniture. It is now a canal information centre, which tells the story of how the hiring of sacks 200 years ago revolutionised the way that corn merchants and

The Wilts & Berks Canal in Swindon town centre in 1914. (Doug Small's collection)

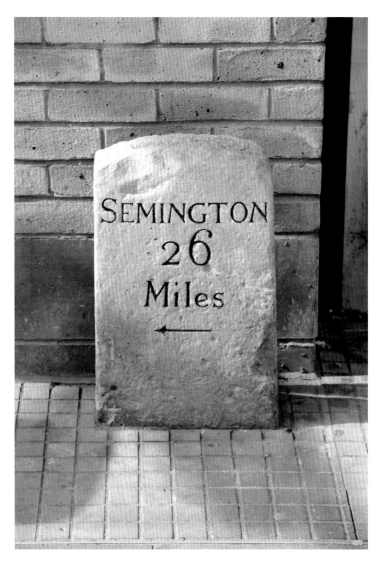

An original milestone from the Wilts & Berks Canal now located in a Swindon shopping centre. (John Minns)

farmers were able to transport their products both in larger quantities and over much longer distances. At the village of Childrey, the historic wharf has been restored and will one day serve as a pleasant visitor mooring location.

South of the village of Shrivenham, the canal had been filled in and an original brick-arch bridge long since removed. Before reopening the waterway, Steppingstone Lane Bridge would have to be re-provided to carry a bridleway across the canal. When the area was excavated in 2003, the original bridge foundations were found to be in good condition and a decision was made to build a new brick-arch bridge on those foundations. The bridge would be built following traditional construction methods by building the arch on temporary timber supports or centring. Since that time, hundreds of hours of navvy work has been provided by volunteers from the Trust and from the Waterway Recovery Group of the Inland Waterways Association to create a bridge of which we are all proud.

But it is at Swindon, the hub of the Wilts & Berks network, that other major blockages have to be resolved. In 1914 the canal still existed within Swindon, and how we wish we were starting our restoration project at this stage of its dereliction! Sadly that was not to be; instead, as the old line approaches the centre of the town, it becomes completely unrecognisable, as it has been completely filled in with little evidence of its existence – every utility company in the area has at one time or another descended on the vacant track to bury their hardware. Everything from sewage pipes, water mains and even modern telecommunications equipment litters the old canal, making it very difficult to just simply dig out the channel. Further on, and into the shopping district at the centre of the town, nothing remains, although one street is named Canal Walk and actually follows the line of the canal. When the shopping development was being built, the architect thought that it would be a good idea to celebrate the history of the place by positioning an historic milestone in Canal Walk.

That milestone indicates that it is only 26 miles (42km) to Semington and the junction with the Kennet & Avon Canal. Unfortunately, it was originally sited on the other side of the pedestrian street from where it sits today and so pointed the wrong way – back towards Abingdon instead of west towards Semington.

Clearly the canal will not be going through this modern town centre development any time soon! So, with the historic north–south route through the heart of Swindon blocked and with no realistic prospect of ever using it again, the Trust commissioned a consultant to find a new route to the west of the town. Then, to the Trust's amazement, the leader of the Town Council called us in, stating, 'this bypass to Swindon you are planning – we don't want it.' We, of course, at first wondered what on earth we had done to upset this man, until he went on to explain that Swindon Borough Council had been inspired by how waterway restoration within other urban environments (like Birmingham) had contributed to the economic regeneration of those environments, and he told us that he wanted the canal to go back right through the centre of Swindon to encourage its regeneration. (Birmingham has more canals than Venice and, like Venice, these canals are now an essential part of Birmingham's growing tourist industry.) The council's leader further added that the restored waterway need not follow precisely the old route through town but that, in places, some existing roads might be dug

Part of the new Wilts & Berks southern bypass canal and housing at Wichelstowe, Swindon, 2011. (John Minns)

up to create tree-lined waterways with wide pedestrian boulevards on either side linking public parks with historic buildings and a new marina that would be created for visiting boats right in the heart of the shopping district. The council published a poster showing an artist's impression of a new waterway, masking an existing busy traffic-choked street as it passed Isambard Kingdom Brunel's Railway Village, a Grade I-listed group of buildings important to the history of Swindon.

North of the proposed visitor marina, the original line of the canal sets off in a north-westerly direction and is, apparently, immediately blocked by the multiple tracks of the main London–Bristol and South Wales railway line. Reinstating this crossing might have been a very expensive project indeed but, fortunately, the original bridge that carried the railway over the canal still survives. Today this 'tunnel' provides a footpath and cycleway under the railway, so it would not be a huge job to reinstate the canal beneath the

Fitting lock gates to a new lock at East Wichelstowe, Swindon, April 2011. (Doug Small)

tracks where the towpath would continue to provide cycle and foot access to the northern suburbs of Swindon. In those northern suburbs, there has been much development activity in recent years, including the provision of new highway infrastructure. True to their vision of the waterway, once again, reaching the heart of Swindon from both north and south, the Borough Highways Department has made provision in the newly built Purton Road Bridge for the future canal and for a cycle route. Built at the same time as the new highway, the bridge cost a fraction of what it would have cost if built after the new roadway was made operational. Since the construction of the bridge, Trust volunteers have cleared the canal along this length and returned it to water.

In 2007, the Trust and its partners, along with another 645 project hopefuls, bid for National Lottery funding to open the canal and a steam railway between the northern suburbs of Swindon and the town of Cricklade in the Cotswold Water Park. Twenty-four of those projects, including our own, received a development grant of £250,000. Only three projects from the original list managed to secure construction funding. The Cricklade Country Way was not one of them but the partnership remains strong and restoration here continues, albeit more slowly than would have been the case with lottery funding. Trust volunteers continue restoring structures along the canal line (like the aqueduct carrying the canal over the River Key), while volunteers of the Swindon & Cricklade Steam Railway are slowly extending their existing track to link the two towns, and colleagues at Great Western Community Forest continue to obtain land between the two routes to plant trees. With the sizeable volume of housing development planned for this area, it is imperative that this 'green corridor' is secured to provide the thousands of future residents of north Swindon with access routes to the countryside.

With the north–south route through Swindon decided, our attention turned to the route to the east where, once again, town

Restored canal at Foxham Way Bridge, Swindon (prior to a Waitrose development that now occupies the right bank of the canal adjacent to the bridge).

development and new roads (one actually having been constructed along the bed of the canal) meant that restoration of this part of the historic main line through Swindon was not a practicable proposition. After considerable deliberation, the chosen solution was to build a southern bypass loop canal. No longer just lines on a plan, at Wichelstowe the bypass canal has been completed together with a new waterside housing development.

Close to the centre of that development is a brand new twenty-first century reinforced concrete lock and nearby a new aqueduct carries the canal across a widened local river. A win–win scenario has been created: Swindon and the Trust get the canal; the housing developer is provided with a new surface-water drainage facility for this low-lying area; and the developer will also enjoy a substantial markup on the sale price of all new houses facing the waterway. West of Swindon on the main line, the next major blockage is at

Father Christmas ready to board *Dragonfly* with a sack of seasonal goodies. (Robert Yeowell)

Recently restored length of rural canal at Studley Grange, south of Swindon, October 2016. (John Minns)

Melksham, but between Swindon town centre and Melksham there has already been a lot of serious restoration activity.

To the south of the town a rural length of the waterway has been fully restored and a new contractor-built bridge has been constructed across that restored length of the historic line of the canal. A short distance further west sits another volunteer-restored original stone-arch bridge. The leading volunteer, a retired stonemason, contrived to fix one of the stones with a sand mortar so that it might later be easily removed, as he has decided that this is where he eventually wants to rest for eternity to watch the boats go by. It is along this length of waterway that volunteers regularly operate the Trust trip boat *Dragonfly*, providing local residents and visitors with a very different view of Swindon.

At Christmas time, the boat, loaded with excited children, stops at Santa's grotto from where he climbs aboard to dispense gifts, mince pies and mulled wine. The Trust was able to take delivery of

its very first trip boat in September 2010 following the exceptional generosity of one of its long-term benefactors, the Underwood Trust; HRH The Duchess of Cornwall, Patron to the Wilts & Berks Canal Trust, formally named the vessel *Dragonfly*. To ensure that access is available for all to enjoy this unique cruising experience, we worked with apprentices at the BMW Group's MINI plant in Swindon to design, build and fit a wheelchair lift into the bow of the boat. BMW undertook this work as an engineering exercise for its apprentices and at no charge to the Trust.

Further west at Studley Grange, a length of waterway was returned to water in the winter of 2016. Work to improve the landscape continued into the spring of 2017 when canal volunteers, working together with the local Wildlife Trust, planted trees on the side of the canal opposite the towpath. In order to control the level of water in this new length, at its western end volunteers had already designed and built a new spill-weir, allowing any excess water in the canal to flow

The rebuilding of lock no. 3 at Seven Locks, near Lyneham, Wiltshire.

into an adjoining stream. A couple of hundred metres from the weir, volunteers have rebuilt a derelict bridge and a lock that now only requires the fitting of gates before boats can once again pass. Gates are not fitted to restored locks until water levels can be maintained, as wooden gates are best kept immersed in water. Beyond the lock there is a short, unrestored length where the local landowner has yet to be convinced of the merits of this amazing community project. At the other side of his land, leading into the town of Royal Wootton Bassett, a mile of canal has been restored and is in use for small boats; a beautiful and level country walk has been provided for the local community, and the wildlife habitat along this once filled in length has been transformed. At the end of this presently isolated section, a

The Peterborough Arms – a vision that will soon be a reality (as visualised in 2015). (Tim Pyatt)

temporary slipway has been provided to encourage the use of boats on the water where the local Sea Scouts now have a safe place to develop their boat-handling skills.

Further west at Tockenham, near to former RAF Lyneham, a flight of locks known as Seven Locks climbs the hillside. Once the damaged and loose brickwork is removed from a derelict lock, not much might appear to remain but, fortunately, the invert, or foundation of the lock, is often found to be in good condition and so, after many hundreds of volunteer hours, a restored structure can rise out of the ground. Two (of seven) locks have now been com-pleted and just await the fitting of gates and the passage of the first boats. (Two done and only five to go!) Lock no. 2 is the next to be restored. It has an existing unclassified road passing right across the chamber, but agreement has already been reached with the high-way authority that the Trust will first build a new bridge beyond the lock to take the repositioned road, after which we will then be able to rebuild the lock chamber and operate the lock without the hindrance of a highway crossing it.

At Dauntsey another lock has been restored. An enchanting length of canal that is abounding with wildlife has been returned

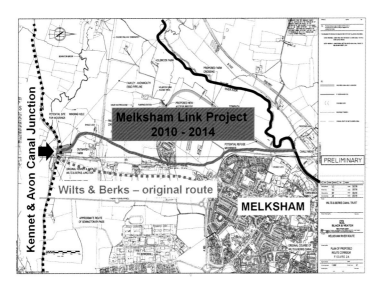

The Melksham Link project joining the Kennet & Avon Canal to the River Avon. (Author; WBCT)

to water and a waterside pub has been acquired by the Trust. Wadworth Brewery, who had owned the pub, closed it in 2014 and applied to the local planning authority for change of use to residential. The Trust, supported by Dauntsey Parish Council and local villagers, blocked the brewery proposal by successfully having the facility declared an 'asset of community value'; the brewery subsequently sold the pub to the Trust. A photo montage of how the Peterborough Arms will look when in full operation gives a good impression of a saved community asset. Volunteers are undertaking restoration work as the availability of funds permits. A community space, lost when the pub closed, has already been reopened. The Trust is saving money, having moved its offices from rented accommodation into the building, and work is on schedule towards the full reopening of the pub.

At Foxham the upper lock has been fully restored, Elm Farm Bascule Bridge is waiting to be raised and Park Farm Lift Bridge (fixed at this time) will have its mechanism installed when required. The area around the head of Calne Branch Canal has been developed by the town council into a town park. The Trust has worked closely with the council to create Castle Fields Park, the new gates of which feature a narrowboat motive within the wrought ironwork. Chaveywell Bridge, located within the park, has been restored by Trust volunteers, and a jetty built nearby supports boat trips that are run at Easter and on other special occasions.

At Pewsham, on the outskirts of Chippenham, are the remains of the most southerly located narrow canal maintenance yard in England. The yard comprises a flight of three locks, the remains of the lock keeper's cottage and an access bridge, together with a dry dock and the remains of workshops – all evidence of a once busy main repair yard of the old canal company. The Trust's current bid to the Heritage Lottery Fund is for the restoration of this historic site, a bid we have jointly submitted in partnership with Chippenham's museum.

Just to the west of Pewsham, now standing majestically in open countryside, is the volunteer-restored Double Bridge. When canals were built across open countryside, landholdings often became divided and affected landowners had the right to demand the construction of accommodation bridges to rejoin their estates or severed farming operations. In this case, instead of building separate bridges for two adjacent landowners, the canny canal engineers saved money by building one shared bridge of extra width, across which the land boundary between the separate estates continued down its centre. The restored bridge was declared open in 2009 by the Trust's patron, HRH The Duchess of Cornwall. The first vessels to pass this way in nearly 100 years were provided enthusiastically by the Wiltshire Youth Canoe Club; nearby, a mechanical digger was removing fill material from the channel as it was slowly working its way towards Melksham and the junction with the Kennet & Avon Canal.

As already stated, Melksham is the site of another major blockage. From its junction with the Kennet & Avon Canal (seen on the left-hand edge of the map above), the original route of the Wilts & Berks

Seated left to right: the author; Michael Lee, Joint Engineering Director, Wilts & Berks Canal Trust; HRH The Duchess of Cornwall. (Tim Pyatt)

Canal (shown with red dots) passed right through the market town of Melksham. The subsequent development of the town has completely blocked this historic line so a new route had to be found. The finally chosen route links the Kennet & Avon Canal to the nearby River Avon (the black line on the right). As chairman of the Trust at the time, I was determined that this project, known as the Melksham Link, should start in 2010 (the 200th anniversary of the opening of the canal) and be completed in 2014, which would be the 100th anniversary of the Act of Parliament that officially abandoned the canal. The project did effectively start in 2010 when HRH The Duchess

of Cornwall cut the first ceremonial sod, a task that all sixty-five children from the nearby primary school enthusiastically assisted her with. Unfortunately, due to the difficulty of fundraising through recent economic recessions, our aim to complete the work by that historic date of 2014 has inevitably slipped somewhat. Detailed plans are, at this time, awaiting final determination by the local planning authority.

The duchess travelled to the sod-cutting ceremony aboard the Trust's narrow-boat *Dragonfly*, affording honorary Engineering Director Mike Lee and myself the opportunity of briefing her on the details of the 3.2km link project that she was, symbolically, about to launch. The Melksham Link will start just to the west of the site of the original junction between the two canals. The original line of the Wilts & Berks Canal used to run through the garden of what is now a private house, which was originally a somewhat unique, three-storey lockkeeper's house; the junction lock is thought to lay beneath the vegetable garden. The arched bridge that once carried the Kennet & Avon towpath over the Wilts & Berks Canal has long since been demolished and the former entrance walled up, with just a trickle of water emanating from a culvert built into the wall.

Most passing boaters on the Kennet & Avon Canal had no idea about the history of the place, and so to draw attention to the former junction, Trust volunteers have recently completed a wonderful piece of artwork on the otherwise blank wall.

Future plans include the development of a new Wilts & Berks marina near to the new junction, which will help to finance the canal that will skirt around the marina and then cut through a disused railway embankment that once carried the former Devizes–Bristol railway line (closed by Dr Beeching in 1966). An historic Bailey bridge, gifted to the Trust by Mendip District Council, will carry a farm track over the canal at this location. How useful, then, that Mike Lee had spent a part of his working life in the British Army putting up and taking down numerous Bailey bridges. (He could not wait, in retirement, to get his hands on this redundant example to play with!) The life-expired bridge has now received hundreds of hours of voluntary restoration work, and, when the component parts are bolted back together again, it will be recycled to cross the canal on a shortened span. Interpretation on the towpath beneath will explain the history of Bailey bridges in general and this bridge in particular – a more fitting end for a unique structure than simply being consigned to a scrapyard.

North of the Bailey bridge, the new waterway will pass through the small village of Berryfield (where a lock will be sited within a landscaped amenity area) before dropping through a further two locks to the River Avon. The navigation will then continue along the river towards Melksham town centre at the existing Melksham Gate Weir. Full navigation standards will be achieved on this short stretch of river by the construction of a new weir across the river just downstream of the canal junction.

Passing beneath Melksham town bridge before reaching the existing weir, the existing unsightly mud banks will be removed as all four arches of this attractive structure are opened up, providing moorings for visiting boats, thus allowing their crews to enjoy the facilities provided by this attractive Wiltshire market town. The existing weir presently creates a 2m-high step change in river levels, although this will reduce to 1.5m once the new weir is in position. A river lock will be required at this location. The plan is to construct a standard Wilts & Berks Canal narrow lock inside a conveniently sited piled area located alongside the weir. What I haven't mentioned yet is that a decision was made some time ago to construct the three locks between the Kennet & Avon and the River Avon to Kennet & Avon broad-gauge standard so that any vessels navigating that waterway will be able to visit the riverside town of Melksham. A short distance up the river from this new lock, the river and the original canal line are reasonably close together; therefore, to complete the bypassing of the blockage of the Wilts & Berks Canal at Melksham, phase two of the Melksham Link project involves cutting a short length of new canal to link across to the historic line.

With the completion of the Wilts & Berks and North Wilts canals (highlighted in yellow on the map earlier in this chapter), two 'cruising rings' will be established: the first composed of the Wilts & Berks, the Thames and the Kennet & Avon and the second the Wilts & Berks, North Wilts, Cotswold Canals and the Thames. Cruising rings are popular with people who rent holiday boats because instead of a there and back return journey, to cruise a ring with ever-changing scenery is much preferred.

So, the impossible dream has gradually turned into an achievable target. It is no longer if but when full restoration will be achieved. Working together with our partners, this canal and all its benefits will in the future be enjoyed by both the people and the wildlife of Wiltshire, Swindon and Oxfordshire, and by the many visitors to the waterway from both home and overseas. I am very proud to say that the Trust's work with local communities was recognised in 2012 when the Trust received the prestigious Queen's Award for Voluntary Service.

If you would like to support this fantastic ongoing restoration project, then you might like to join the charitable Wilts & Berks Canal Trust. The individual annual membership fee is only £10 per year and for that you will receive four magazines (either by post or online) to keep you updated on the progress we are making and the support given to us by our royal patrons. More important to us than the membership fee paid to the charity (although that still buys a bag and a half of cement) is the evidence that an expanding membership provides to the National Lottery Fund of increasing popular support for the project – something they always look for before making their awards. So please consider helping us to win a new lottery grant – your support will be greatly appreciated by the hardworking volunteers of the Trust. Thank you.

www.wbct.org.uk

The bricked-up former junction of the Wilts & Berks and Kennet & Avon canals at Semington, Wiltshire. (Paul Lenearts)